Land Use–Transport Interaction Models

Land Use–Transport Interaction Models

by
Rubén Cordera
Ángel Ibeas
Luigi dell'Olio
Borja Alonso

CRC Press
Taylor & Francis Group
Boca Raton London New York

CRC Press is an imprint of the
Taylor & Francis Group, an **informa** business

CRC Press
Taylor & Francis Group
6000 Broken Sound Parkway NW, Suite 300
Boca Raton, FL 33487-2742

First issued in paperback 2019

© 2018 by Taylor & Francis Group, LLC
CRC Press is an imprint of Taylor & Francis Group, an Informa business

No claim to original U.S. Government works

ISBN-13: 978-1-138-03246-0 (hbk)
ISBN-13: 978-0-368-88455-0 (pbk)

Visit the Taylor & Francis Web site at
http://www.taylorandfrancis.com

and the CRC Press Web site at
http://www.crcpress.com

Contents

SECTION I State of the Art in Land Use–Transport Interaction Modelling

SECTION II The Theoretical Basis of Land Use–Transport Interaction Models

SECTION III   Steps for Building an Operational Land Use–Transport Interaction Model

SECTION IV   Land Use–Transport Interaction Models Considering Spatial Dependence

Acknowledgements

The authors of this book, who are part of the Transport Systems Research Group (Grupo de Investigación de Sistemas de Transporte [GIST]) based at the University of Cantabria, Cantabria, Spain, acknowledge all the members of the group who have directly or indirectly contributed towards its creation. We also thank Professor Pierluigi Coppola of the University of Rome Tor Vergata, Rome, Italy for his inestimable help in obtaining the results on which this research was based and to Chris Tyas, geographer, for his contributions in the review and translation of this book. The active collaboration of Santander City Council, Cantabria, Spain, provided the information required to study the territorial realities used in this research. Finally, we thank the Spanish Ministry of the Economy and Competitiveness and the European Commission for their financial support that made this book a reality, through the following projects: (1) INTERLAND (E 21/08), (2) TRANSPACE (TRA2012-37659), (3) PARK-INFO (TRA2013-48116-R) and (4) SETA: A ubiquitous data and service ecosystem for better metropolitan mobility.

Editors

Dr. Rubén Cordera is a researcher in transport and land use interaction in the Transportation Systems Research Group at the University of Cantabria, Cantabria, Spain, from where he earned his PhD. His research focuses on land use–transport interaction (LUTI) models, transport planning and spatial-econometric models that are applied to transport and land use.

Dr. Ángel Ibeas is a professor of transport planning and director of the Transport Systems Research Group at the University of Cantabria, Cantabria, Spain, from where he earned his PhD. His research focuses on logistics, transit operation, travel behaviour and intelligent transportation systems.

Dr. Luigi dell'Olio is a professor of transport planning and the head of the Transportation Demand Modelling Division of the Transport Systems Research Group at the University of Cantabria, Cantabria, Spain. He earned his PhD from the University of Burgos, Burgos, Spain and his master of engineering from the Technical University of Bari, Bari, Italy. His research focuses on transit operations and planning, travel behaviour and land use and transport interaction models.

Dr. Borja Alonso is an assistant professor of transport planning at the University of Cantabria, Cantabria, Spain, from where he earned his PhD. He belongs to the Transport Systems Research Group since 2007 and his research focuses on transit operations and planning, transport modelling and traffic engineering.

Contributors

Gonzalo Antolin
University of Cantabria
Cantabria, Spain

Juan Benavente
University of Cantabria
Cantabria, Spain

Eneko Echániz
University of Cantabria
Cantabria, Spain

Sara Ezquerro
University of Cantabria
Cantabria, Spain

Hugo González
University of Cantabria
Cantabria, Spain

Alvaro Landeras
University of Cantabria
Cantabria, Spain

Introduction

AN INTRODUCTION TO LAND USE–TRANSPORT INTERACTION MODELS

The main reason for writing this book is to provide a theoretical as well as an applied introduction to operational interaction models for land use and transport. The interdependent relationship between the land use and transport is widely recognised in the theoretical field and through applied experience. The configuration of transport systems, made up of networks and flows, is heavily influenced by the distribution and the characteristics of land use, be it residential, commercial or others. The demand for transport comes from the need that people have to perform activities that are physically separated in space. If, in an ideal world, all land uses and activities could coincide at one point, then the demand for transport would be inexistent. It is the need to move generated by the desire to carry out different activities at different physical installations that explain the demand for transport (Rodrigue et al. 2013). Transport can also condition the distribution, characteristics and growth of different land uses, given that it modifies the accessibility conditions of different places making up the urban space. It would appear reasonable to suppose that a highly accessible location well served by the transport links would be a more attractive location for companies to install commercial premises. Other highly accessible areas for different types of opportunities (e.g. education, leisure) could be more attractive for the development of other kinds of land use. So, changes made to the transport system can have an influence on the land use system and vice versa, until equilibrium is reached between the two. The urban system as a whole (a small city, a large city and even an urban region) can be understood through simplified land use–transport interaction (LUTI) modelling of aspects such as the interaction between the transport and land use sub-systems, as shown in Figure I.1.

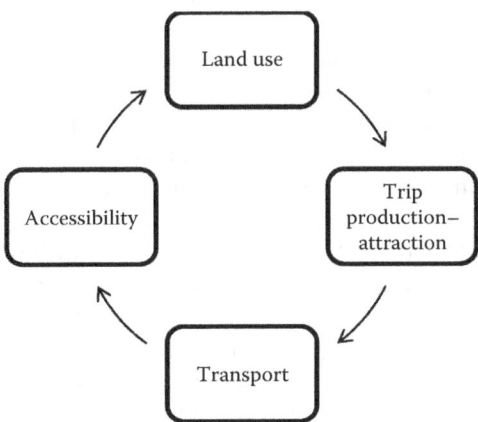

FIGURE I.1 Interaction between the transport and land use sub-systems.

Although the conceptualisation of transport as a sub-system formed by the physical supply of transport services and a demand for travel are supported by a wide consensus in the literature, usage of the term land use, as highlighted by Goulias (2003) is more complex and has generally been applied for different contexts grouping processes as diverse as

- Land development, or the changes made to the urban form and the built space.
- Location choices made by urban agents, mainly households and companies.
- The spatial pattern of activities through which households live their daily lives: home, work, consume.
- The physical interaction of goods and services between companies and between the companies and households.

As an analogy with the conceptualisation of transport, the land use sub-system may be divided into a supply made up of built space (physical installations such as housing, commercial buildings, etc.) and a demand formed by the location choices made by the urban agents during their daily activities.

The models presented in this book are abstract models providing a simplified representation of the complex reality of urban systems through the use of mathematical equations. The different theoretical bases that form the nucleus of LUTI models try to capture the fundamental relationships determining the logic behind the urban systems and their response to different actions. These actions are generally related to urban policies and transport issues such as new housing developments, new transport infrastructure or the introduction of new equipment. LUTI models try to respond to common questions that emerge from research or practical application in the field of planning (Martínez 2000): Under what circumstances do transport projects induce urban development? Up to what point are location decisions made by the urban agents influenced by the accessibility of places? How will the future growth of an urban area be distributed? What proportion of the benefits generated by a transport project increasing location accessibility is taken advantage of by land owners and users?

LUTI models have traditionally been subjected to similar criticism to that directed at the mathematical models used in the social sciences and especially in urban and transport planning. Some of these criticisms have highlighted aspects such as the need to collect large amount of data to feed the models with the inherent complexity of the models, which sometimes makes them work similar to an authentic *black box*, the need to program complicated algorithms, the high cost of calibration and maintenance of the models and their inability to simulate highly complex systems, for example, urban systems, leading to significantly erroneous predictions. Contrary arguments have also appeared defending the use of models in researching and planning urban systems, such as the need to use a coherent theoretical and explicit framework for analysing the repercussions of different actions and the support that the models provided by imposing a certain discipline around a more conscious and precise understanding about the workings of the reality being studied. Furthermore, models are becoming increasingly cheaper to use and provide greater utility; thanks to the generalisation of easily accessible digital data, the continuing improvements in computing speed and more efficient and better programming techniques. Obviously, this book considers the latter arguments to be of greater weight and will encourage the continuing research and application of mathematical models, which allow us to understand and better predict the repercussions of applying different policies in urban systems. Nevertheless, the reader should be aware that a LUTI model is a tool to be applied by a professional or a team of professionals who should understand their potential as well as their limitations if they are to be correctly used and the results precisely interpreted. The intelligence of the model will definitely not overtake the intelligence of the human group in charge of feeding it, running it and interpreting its results.

ORGANISATION OF THE BOOK

This book is structured into four large sections. The first of them: Section I 'State of the Art in Land Use–Transport Interaction Modelling' provides a history of LUTI modelling since their origin in the United States in the 1960s. This section also provides a brief description and classification of some of the better known models that are applied in different areas and study cases, which is why they have been denominated as operational models rather than theoretical models developed purely in an academic research setting. Section II 'The Theoretical Basis of Land Use–Transport Interaction Models' summarises the main concepts and theories, which have been and are used as the basis of LUTI models trying to simulate the behaviour and balance of urban systems. The central concept of accessibility is described as the authentic nexus of the union between the transport and land use. This is followed by a summary of the most important theories that have surfaced to spatially explain urban systems within the field of LUTI modelling: urban microeconomic theory, spatial interaction theory, random utility theory and optimisation theory. Section III 'Steps for Building an Operational Land Use–Transport Interaction Model' introduces the practical aspects involved in the specification and implementation of a LUTI model. Among these aspects is the fundamental point of defining the model's goal in order to use a system, which truly represents the detail, complexity and phenomena that are modelled to address the practical needs of the modeller. Other basic aspects involved in designing the implementation of a model are related to the characteristics of the study area, the required level of detail and the available data, which, depending on the study area, could be insufficient for certain practical ends. Section III also describes the main submodels that normally form part of a LUTI system: models for the location of the population and activities, models simulating the impact of transport on real estate values and models for simulating the transport system, typically summarised into the following four classic stages: (1) trip generation, (2) trip distribution, (3) modal distribution and (4) network assignment. These sections are illustrated through the use of examples provided by a specific urban system, the city of Santander, a medium-sized city located on the North coast of Spain. Finally, Section IV 'Land Use–Transport Interaction Models Considering Spatial Dependence' introduces more novel contributions in the field of modelling in which spatial dependence in the data is considered. Given the intense usage of strongly spatial data in LUTI models, it is important to consider the possible presence of spatial dependence, that is, the dependence between observations or between choice alternatives derived from their greater similarity to the phenomena present in other points in space. This spatial dependency is based on what is known as Tobler's first law of geography 'everything is related to everything else, but near things are more related than distant things' (Tobler 1970). The consideration of this concept is of relevance when estimating models fulfilling the statistical properties from which econometric techniques such as multiple linear regression or discrete choice come. A more detailed description of some of the more frequently used models in the field of LUTI modelling adapted to the possible presence of spatial correlation in the data is provided in Section IV: population and activity location models, hedonic regression models and trip generation models.

This book is targeted at the academic and the professional worlds interested in the field of urban and transport modelling. Given the mathematical nature of the models being presented, a certain degree of previous knowledge is required to fully understand all the results. However, the text is of an introductory nature, so the models are relatively simple and a basic knowledge of algebra and statistics should be enough to understand them. Further reading is recommended and other books, which will complement the present text on specific techniques such as discrete choice (Ben-Akiva and Lerman 1985, Train 2009, Hensher et al. 2015), multiple linear regression (Gujarati and Porter 2009), transport modelling (Cascetta 2009, Ortúzar and Willumsen 2011) or general urban planning (Weber and Crane 2012) are useful. Older references that the reader may find valuable in understanding the field of LUTI modelling are the works of Foot (1981) and Barra (1989). Both books are slightly out of date but they are still useful in providing an introduction to modelling the interaction between transport and land use.

REFERENCES

Barra, T. 1989. *Integrated Land Use and Transport Modelling: Decision Chains and Hierarchies, Cambridge Urban and Architectural Studies.* Cambridge, UK: Cambridge University Press.

Ben-Akiva, M. E. and Lerman, S. R. 1985. *Discrete Choice Analysis: Theory and Application to Travel Demand*, Vol. 9. Cambridge, MA: MIT Press.

Cascetta, E. 2009. *Transportation Systems Analysis: Models and Applications*, 2nd ed. *Springer Optimization and Its Applications.* New York: Springer.

Foot, D. H. S. 1981. *Operational Urban Models: An Introduction.* London, UK: Methuen.

Goulias, K. G. 2003. *Transportation Systems Planning: Methods and Applications, New Directions in Civil Engineering.* Boca Raton, FL: CRC Press.

Gujarati, D. N. and Porter, D. C. 2009. *Basic Econometrics*, 5th ed. Boston, MA: McGraw-Hill Irwin.

Hensher, D. A., Rose, J. M. and Greene, W. H. 2015. *Applied Choice Analysis,* 2nd ed. Cambridge, UK: Cambridge University Press.

Martínez, F. J. 2000. Towards a land-use and transport interaction framework. In *Handbook of Transport Modelling*, Hensher, D. A. and Button, K. J. (Ed.), pp. 145–164. Amsterdam, the Netherlands: Elsevier Science.

Ortúzar, J. D. and Willumsen, L. G. 2011. *Modelling Transport*. Hoboken, NJ: John Wiley & Sons.

Rodrigue, J. P., Comtois, C. and Slack, B. 2013. *The Geography of Transport Systems*. Boca Raton, FL: Taylor & Francis Group.

Tobler, W. R. 1970. A computer movie simulating urban growth in the detroit region. *Economic Geography* 46:234–240. doi:10.2307/143141.

Train, K. 2009. *Discrete Choice Methods with Simulation*, 2nd ed. Cambridge, UK: Cambridge University Press.

Weber, R. and Crane, R. 2012. *The Oxford Handbook of Urban Planning*. Oxford, UK: Oxford University Press.

Section I

State of the Art in Land Use–Transport Interaction Modelling

This section provides an up to date state-of-the-art review of land use–transport interaction (LUTI) modelling, including the most recent developments found in the literature. A brief history of LUTI modelling shows that it coincided with the surge in transport modelling occurring in the United States and the application of quantitative methods in multiple fields of social sciences. This section provides a classification of various implemented operational LUTI models that have been used for planning purposes in different study areas. This classification could serve as a guide for the reader when choosing the appropriate LUTI model for their particular requirements.

1 A Brief History of Land Use–Transport Interaction Models

Rubén Cordera, Ángel Ibeas, Luigi dell'Olio and Borja Alonso

The modelling of urban systems based on the simple principle that there is an interaction between land use and transport goes back to the late 1950s and the early 1960s. Nevertheless, several theoretical advances had been made before that, which were relevant to the field of spatial location and spatial interaction. The work of Reilly (1931) in estimating the market areas of nearby cities considering their population sizes, and the location theories of the German school of von Thünen (1826), Christaller (1933) and Lösch (1954) (see also Chapter 4) on the distribution of land uses and central places, made important contributions towards the mathematical modelling of the phenomena that is related to the spatial location of populations and economic activities.

It is towards the end of the 1950s in the United States of America when various convergent phenomena occurred to explain the surge in land use–transport interaction (LUTI) modelling (Putman 1983). First, the demand from the administrations for trustworthy theoretical estimations about what impact the investments made in motorway construction would have on land use was the key. Second, the increasing urban problems and a greater awareness of the need for technical knowledge in order to address them added to the demand for accurate models. Finally, the greater availability of data and digital equipment increased the possibilities of mathematical modelling and simulation. This expansion in the field of LUTI modelling occurred somewhat behind the development of research in transport modelling; a rapidly growing field connected to growth in the American cities and increased rates of motorization derived from the post-war economic boom. In the fields of transport and urban planning, it soon became clear that there was a need to consider the relationship between transport and land use as endogenous, given the importance of the spatial distribution of activities in understanding the patterns of trip generation and distribution. In this sense, the work of Hansen (1959) was pioneering because it empirically examined how residential developments in the urban area of Washington DC were related to the available space and the accessibility by road to different opportunities. Hansen based his research on previous theoretical work into social physics such as Zipf (1949) and Stewart (1948) to formulate an accessibility indicator based on a gravitational analogy and defining accessibility as a measurement of the possibility of realising an interaction. Later, Huff (1963) followed on the work of

Reilly to interpret the gravity model in both probabilistic and utility terms by providing a formula similar to a discrete choice logit model and thereby equipped it with a greater theoretical content.

Based on this research, Lowry (1964) developed what is now considered as the first operational LUTI model (Figure 1.1). Lowry combined the theory of spatial interaction with economic base theory (Hoyt 1941), which states that two economic sectors exist in an urban system: (1) a basic one where location is exogenous to the model and (2) another non-basic one where the distribution depends on the location of the residents in the study area. In turn, the location of the population depends on the location of their place of work to provide location equilibrium between the non-basic economic sector and the residential sector. The main contributions made by this model were, therefore, the combination of an economic theory with a spatial theory to formulate an operationally balanced iterative model for making planning predictions. The Lowry model was, in fact, implemented in various American cities during the 1960s, most notably in Pittsburgh where it was originally developed. Experiments were also carried out in Boston using linear regression techniques (Lowry 1967) and in Pennsylvania, one of the main research centres in urban modelling due to the theoretical work of Alonso (1964) and other more applied work with the development of the Penn–Jersey model (Herbert and Stevens 1960) (Chapter 7). These models were normally developed as part of urban planning work to evaluate the evolution of different policies. The Lowry model was later modified by Garin (1966) who added a matricial formulation, which allowed a direct solution without the need for iterations. Other models were also developed in the United States following the structure developed by the Lowry model, such as the time-oriented metropolitan model (TOMM) developed by Crecine (1964) and the projective land use model (PLUM) adding residential location and commercial activity models through the use of probabilistic formulation (Putman 1979).

Urban modelling spread relatively quickly from the United States to the United Kingdom towards the end of the 1960s and the start of the 1970s. The models developed in the United Kingdom were largely based on extensions of the Lowry model and had a smaller scale and lower cost than those that were developed in the United States (Foot 1981). Some of the first active research groups were based at the Universities of Reading, Cambridge, Leeds and Centre for Advanced Spatial Analysis (CASA) at the University College London (UCL) (Lun and Ying 2015). Echenique et al. (1969) introduced a real estate supply location sub-model into the Lowry model, together with an activities location sub-model to consider the relationship between the supply and the demand of location. The introduction of maximum entropy theory by Wilson (1970) allowed the theory of spatial interaction to be linked with location models within a consistent theoretical framework based on the techniques of optimising and restrictions to origins or destinations.

At the same time as these developments were occurring in the United Kingdom and other countries at the beginning of the 1970s, a critical movement began to show scepticism towards LUTI models. A famous article written by Lee (1973) criticised LUTI models because they did not fulfil the expected goals and diverted attention towards the use of alternative methods, which were, following Lee, simpler and more efficient in answering socially useful questions. Lee summarised the criticism

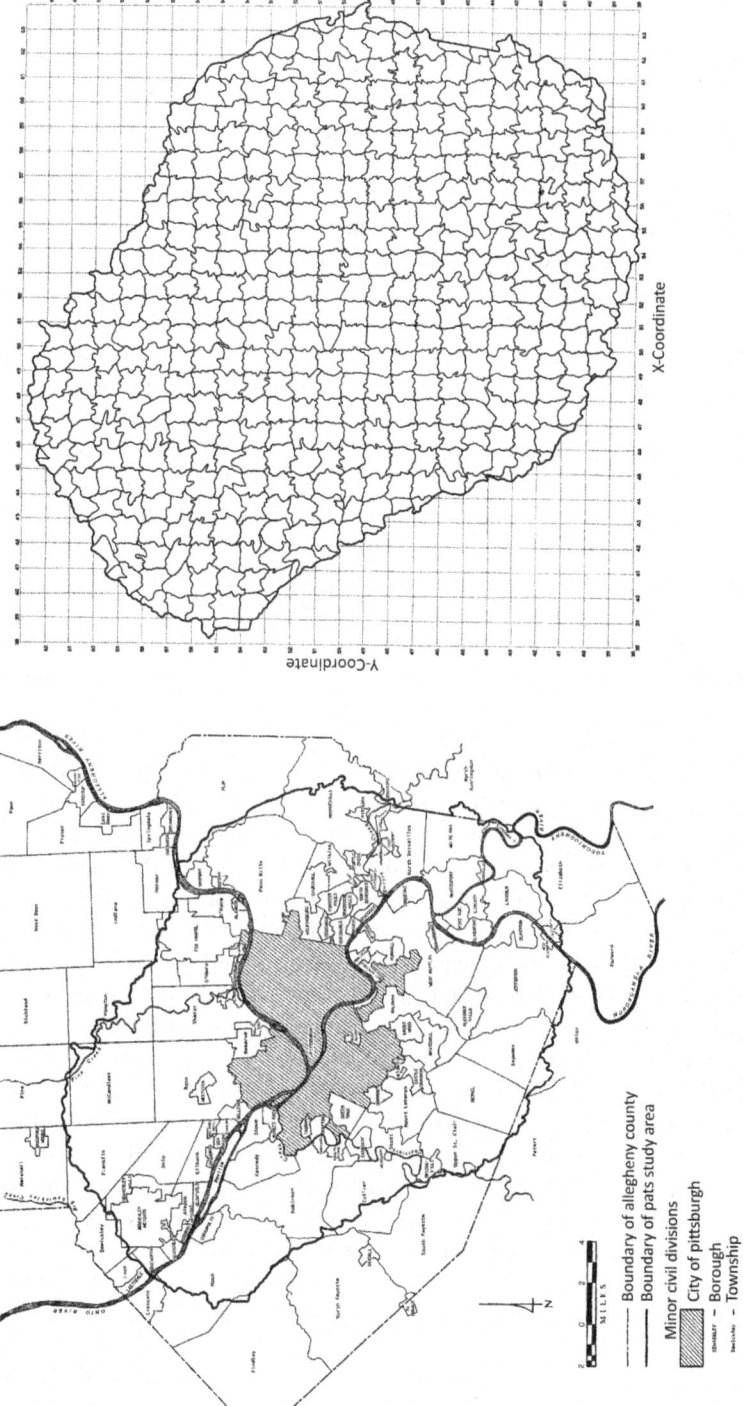

FIGURE 1.1 Study area and zoning used in the Pittsburgh Lowry (1964). Note: Tracts are identified by coordinates of the lower left corner.

towards LUTI models into seven large problems: (1) the need to evaluate an excessive number of planning objectives using only one technique, (2) the excessive aggregation of simulation results, (3) the need for a large volume of data in order for the models to produce results, (4) predictions that are quite often clearly wrong, (5) the excessive functional complexity, (6) the excessive dependency on software which, for the planner, behaves similar to a black box and (7) the high cost of implementation. Instead, Lee suggested greater equilibrium between theory, data and intuition, a greater focus on real urban problems and the use of simpler models that were adapted to the problems of interest. In this sense, the financing and the high expectations placed on the models were significantly reduced in order to confront more immediate problems related to urban poverty and inequality for which LUTI models were less useful (Batty 1976). Planning styles also changed from being more centralised to more incremental and focussed. However, although these problems reduced the expectations, they did not eliminate research into simulating urban reality nor in application. Planning agencies in the United States and other places continued to use LUTI models, especially as evaluation tools providing different measurements to support strategic plans at an urban or regional scale (Pack and Pack 1977).

Since the end of the 1970s and during the 1980s research into the use of LUTI models continued to develop along different lines of investigation. In Great Britain, Echenique et al. (1974) incorporated a more disaggregated sub-model into the Lowry model, which integrated maximisation of entropy with microeconomic theory. This group also performed different practical applications not only in Europe but also in the Latin America. The work carried out in the University of Pennsylvania (Putman 1983) culminated in the development of the integrated transportation and land use package model (ITLUP) incorporating different extensions with respect to the Lowry model and breaking down the structure into two large modules: (1) a residential location model (DRAM) and (2) an employment location model (EMPAL). The ITLUP model incorporated a greater degree of disaggregation along with transport modal choice and assignment sub-models. It needs to be noted that this model comes with various modifications and thanks to its lower data requirements and its adaptation to planning needs that is still being extensively used in the United States and other countries (Porter et al. 1996, Iacono et al. 2008). Batty (1976), working at UCL, also performed important work consolidating the available knowledge about LUTI modelling and various applications in the United Kingdom.

It can reasonably be said, therefore, that the 1970s and 1980s were a period of consolidation in the field of urban modelling. In spite of the problems within the research field and the excessive optimism placed on urban models during the 1960s, the techniques used, the available theory and the field applications continued to grow progressively. All this was thanks to the continuous work of various research groups located in the United States and Europe and new research starting in other countries, most notably Japan (Aoyama 1989). Advances were made to address problems associated with model estimation and calibration techniques as well as in the required scale of zoning. Furthermore, the increasing amount of accumulated knowledge started to be presented in the first introductory textbooks for urban modelling (Foot 1981, Batty 1976, de la Barra 1989). In the 1980s, very significant advances were also made in parallel fields such as information technology (IT) and digitalised data, which endowed the LUTI models with greater power. The systems for managing alphanumerical data, geographical data (GIS) and other software such as computer assisted design (CAD), spreadsheets and expert systems were clear examples of this development. During this decade, the more active British research groups were located in Liverpool, Leeds and above all Cambridge through the Martin Centre for Architectural and Urban Studies led by Echenique (Echenique and Owers 1994). This group used the MEPLAN model to perform practical applications in London and South East England (Williams 1994), the Channel Tunnel (Rohr and Williams 1994), Italy (Hunt 1994), Spain (Burgos 1994) and Latin America (de la Barra 1994). The International Study Group on Land Use/Transport Interaction (ISGLUTI) was also formed in the 1980s with the goal of comparing different land use and transport interaction models. The first phase of the ISGLUTI work compared different LUTI models such as TOPAZ, DORTMUND, MEPLAN, LILT and others to check the consistency of their results. The results were a little disappointing because of the great differences in data requirements of the different models (Webster et al. 1988). The second phase of the ISGLUTI work applied various LUTI models (CALUTAS, DORTMUND, LILT and MEPLAN) to the same study areas at the same time where more than one model was being applied to different cities. These models were used to simulate the results of different socioeconomic dynamics and policies (population change, employment change, changes in commercial policies, among others) and concluded that the results of different models were consistent to an aggregated level, but less so on a disaggregated level (Paulley and Webster 1991).

From the 1990s onwards, the number of available LUTI models and their application to real cases continued to rise. Some authors have called this decade the golden age of LUTI modelling (Jones 2016). Wegener (1994) identified 20 research centres where intense activity in LUTI modelling was taking place to evaluate different policies. Seven of them were located in the United States, six in Europe, four in Asia, two in Latin America and one in Australia. Wegener (2004) also reviewed a large part of the models that were available in the literature and chose 20 LUTI models based on different modelling techniques. Furthermore, the range of policies being evaluated by the models saw an increase derived from the greater interest in knowing about the effects of transport on land use and the effect of both subsystems on the environment. The SPARTACUS and PROPOLIS research projects (Lautso et al. 2004), financed by the European Commission, evaluated different scenarios in various study areas to simulate the effects of the relationship of transport and land use in terms of pollution, social impacts on health and economic well-being in general. Different legislation was passed in the United States such as the Intermodal Transportation Efficiency Act (ISTEA) (Gage and McDowell 1995), which also encouraged the use of LUTI models for evaluating the repercussions of infrastructure not only on the transport system but also on land use and the environment.

Research and practical applications using LUTI models are currently in a good state of health. Some of the subjects that are currently being addressed by research are: microsimulation, in other words, the use of high-resolution models with greater data requirements (Wegener 2011); the integration of LUTI models and activity-based models (Acheampong and Silva 2015); the use of LUTI models to evaluate environmental impacts; the simulation of the complete dynamics of urban systems instead of their instant transition into points of equilibrium (Simmonds et al. 2013); the application of spatial decision models or the interrelationship with other modelling paradigms such as cellular automata or agent-based models (Timmermans 2003); and the application of spatial-econometric techniques (dell'Olio et al. 2016). The developments made for practical applications concentrate on making the models more operational, increasing their standardisation in both and the formatting of the data and the software, in order to increase their operability and their capacity to adapt for the evaluation of different problems (Hardy 2012, Saujot et al. 2015).

REFERENCES

Acheampong, R. A. and Silva, E. 2015. Land use–transport interaction modeling: A review of the literature and future research directions. *Journal of Transport and Land Use* 8 (3):11–38.

Alonso, W. 1964. *Location and Land Use: Toward a General Theory of Land Rent, Publications of the Joint Center for Urban Studies of the Massachusetts Institute of Technology and Harvard University.* Cambridge, UK: Harvard University Press.

Aoyama, Y. 1989. A historical review of transport and land-use models in Japan. *Transportation Research Part A: General* 23 (1):53–61. doi:10.1016/0191-2607(89)90140-4.

Batty, M. 1976. *Urban Modelling: Algorithms Calibrations, Predictions, Cambridge Urban and Architectural Studies 3.* Cambridge, UK: Cambridge University Press.

Burgos, J. L. 1994. Integrated land-use and transport models in the Basque Country. *Environment and Planning B: Planning and Design* 21 (5):603–610. doi:10.1068/b210603.

Crecine, J. P. 1964. *A Time-Oriented Metropolitan Model for Spatial Location.* Pittsburgh, PA: Community Renewal Program, Department of City Planning.

Christaller, W. 1933. *Die zentralen Orte in Süddeutschland.* Jena, Germany: Gustaf Fisher. Translated by Carlisle, W. B. (1966), as *Central Places in Southern Germany.* Englewood Cliffs, NJ: Prentice Hall.

de la Barra, T. 1989. *Integrated Land Use and Transport Modelling: Decision Chains and Hierarchies, Cambridge Urban and Architectural Studies.* Cambridge, UK: Cambridge University Press.

de la Barra, T. 1994. From theory to practice: The experience in Venezuela. *Environment and Planning B: Planning and Design* 21:611–617.

dell'Olio, L., Ibeas, Á. and Cordera, R. 2016. Project TRANSPACE - Advanced land use and transport interaction model (TRA2012-37659). Ministry of Economy and Competitiveness (Goverment of Spain).

Echenique, M., Crowther, D. and Lindsay, W. 1969. A spatial model of urban stock and activity. *Regional Studies* 3 (3):281–312.

Echenique, M., Feo, A., Herrera, R. and Riquezes, J. 1974. A disaggregated model of urban spatial structure: Theoretical framework. *Environment and Planning A* 6 (1):33–63.

Echenique, M. and Owers, J. 1994. Research into practice: The work of the martin centre in urban and regional modelling. *Environment and Planning B: Planning and Design* 21 (5):513–515. doi:10.1068/b210513.

Foot, D. H. S. 1981. *Operational Urban Models: An Introduction.* London, UK: Methuen.

Gage, R. W. and McDowell, B. D. 1995. ISTEA and the role of MPOs in the new transportation environment: A midterm assessment. *Publius: The Journal of Federalism* 25 (3):133–154.

Garin, R. A. 1966. A matrix formulation of the lowry model for intrametropolitan activity allocation. *Journal of the American Institute of Planners* 32 (6):361–364. doi:10.1080/01944366608978511.

Hansen, W. G. 1959. How accessibility shapes land use. *Journal of the American Institute of Planners* 25 (2):73–76.

Hardy, M. 2012. Using open source data to populate, calibrate and validate a simplified integrated transportation and land use model. *Transportation Research Board 91st Annual Meeting*, Washington, DC, January 22–26.

Herbert, J. D. and B. H. Stevens. 1960. A model for the distribution of residential activity in urban areas. *Journal of Regional Science* 2 (2):21–36. doi:10.1111/j.1467-9787.1960.tb00838.x.

Hoyt, H. 1941. Economic background of cities. *The Journal of Land & Public Utility Economics* 17 (2):188–195.

Huff, D. L. 1963. A probabilistic analysis of shopping center trade areas. *Land Economics* 39 (1):81–90. doi:10.2307/3144521.

Hunt, J. D. 1994. Calibrating the naples land-use and transport model. *Environment and Planning B: Planning and Design* 21 (5):569–590. doi:10.1068/b210569.

Iacono, M., Levinson, D. and El-Geneidy, A. 2008. Models of transportation and land use change: A guide to the territory. *Journal of Planning Literature* 22 (4):323–340.

Jones, J. 2016. Spatial bias in LUTI models. PhD Thesis, Université catholique de Louvain, Louvain-la-Neuve, Belgium.

Lautso, K., Spiekermann, K., Wegener, M., Sheppard, I., Steadman, P., Martino, R., Domingo, A. and Gayda, S. 2004. PROPOLIS: Planning and research of policies for land use and transport for increasing urban sustainability, Final report. Project Funded by the European Commission.

Lee, D. B. 1973. Requiem for large-scale models. *Journal of the American Planning Association* 39 (3):163–178.

Lösch, A. 1954. *The Economics of Location*. New Haven, CN: Yale University Press.

Lowry, I. S. 1967. *Seven Models of Urban Development: A Structural Comparison*, Vol. 3673. Santa Monica, CA: Rand Corporation.

Lowry, I. S. 1964. *A Model of Metropolis, Memorandum*. Santa Monica, CA: Rand Corporation.

Lun, L. and Ying, L. 2015. A retrospect and prospect of urban models: Reflections after interviewing Michael Batty. *China City Planning Review* 24 (4):63–70.

Pack, H. and Pack, J. R. 1977. The resurrection of the urban development model. *Policy Analysis* 3 (3):407–427.

Paulley, N. J. and Webster, F. V. 1991. Overview of an international study to compare models and evaluate land-use and transport policies. *Transport Reviews* 11 (3):197–222.

Porter, C., Melendy, L. and Deakin, E. 1996. *Land Use and Travel Survey Data: A Survey of the Metropolitan Planning Organizations of the 35 Largest U.S. Metropolitan Areas*. Berkeley, CA: Institute of Urban and Regional Development.

Putman, S. H. 1979. *Urban Residential Location Models, Studies in Applied Regional Science*, Vol. 13. Boston, MA: Martinus Nijhoff Publishers.

Putman, S. H. 1983. *Integrated Urban Models*. 2 vols, Vol. 1: *Research in Planning and Design*. London, UK: Pion.

Reilly, W. J. 1931. *The Law of Retail Gravitation*. New York: W.J. Reilly.

Rohr, C. and Williams, I. N. 1994. Modelling the regional economic impacts of the channel tunnel. *Environment and Planning B: Planning and Design* 21 (5):555–567. doi:10.1068/b210555.

Saujot, M., De Lapparent, M., Arnaud, E. and Prados, E. 2015. To make LUTI models operationnal tools for planning. *International Conference on Computers in Urban Planning and Urban Management (CUPUM)*, Cambridge, MA, July 7–10.

Simmonds, D., Waddell, P. and Wegener, M. 2013. Equilibrium versus dynamics in urban modelling. *Environment and Planning B: Planning and Design* 40 (6):1051–1070. doi:10.1068/b38208.

Stewart, J. Q. 1948. Demographic gravitation: Evidence and applications. *Sociometry* 11 (1/2):31–58. doi:10.2307/2785468.

Timmermans, H. 2003. The saga of integrated land use-transport modeling: How many more dreams before we wake up. Keynote paper, moving through nets: The physical and social dimension of travel. *10th International Conference on Travel Behaviour Research*, Lucerna, Switzerland.

von Thünen, J. H. 1826. *Der isolierte staat in beziehung auf landwirtschaft und nationaloekonomie*. Jena, Germany: Gustaf Fisher. Translated by Wartenburg, C. M. (1966) as *The Isolated State*. Oxford, UK: Oxford University Press.

Webster, F. V., Bly, P. H., Paulley, N. J., Brotchie, J. F. and International Study Group on Land-Use/Transport Interaction. 1988. *Urban Land-Use and Transport Interaction: Policies and Models: Report of the International Study Group on Land-Use/Transport Interaction (ISGLUTI)*. Aldershot, UK: Avebury.

Wegener, M. 1994. Operational urban models state of the art. *Journal of the American Planning Association* 60 (1):17–29.

Wegener, M. 2004. Overview of land-use transport models. In *Transport Geography and Spatial Systems*, David, A.H. and Kenneth, B. (Ed.), pp. 127–146. Kidlington, UK: Elsevier.

Wegener, M. 2011. From macro to micro-how much micro is too much? *Transport Reviews* 31 (2):161–177.

Wilson, A. G. 1970. *Entropy in Urban and Regional Modelling, Monographs in Spatial and Environmental Systems Analysis 1*. London, UK: Pion.

Williams, I. N. 1994. A model of london and the south east. *Environment and Planning B: Planning and Design* 21 (5):535–553. doi:10.1068/b210535.

Zipf, G. K. 1949. *Human Behavior and the Principle of Least Effort*. Cambridge, MA: Addison-Wesley.

2 A Classification of Land Use–Transport Interaction Models

Rubén Cordera and Ángel Ibeas

CONTENTS

A large number of land use–transport interaction (LUTI) models are currently available in the literature for research and practical planning purposes. Different authors have proposed classifications, which group different models according to different criteria. Section 2.1 gives an overview of the criteria used by these authors to classify LUTI models, followed by Section 2.2 that provides a review of the most commonly known LUTI models that have been applied in a greater number of study areas. Finally, Section 2.3 proposes a new classification of LUTI models based on those described beforehand but which combines the main theoretical paradigms developed in the field of LUTI modelling with their chronological evolution.

2.1 CRITERIA FOR CLASSIFYING LAND USE AND TRANSPORT INTERACTION MODELS

The complexity of LUTI models allows them to be classified using multiple criteria. Wegener (2004) has proposed nine main criteria, which can be used to classify these kinds of models:

1. *Comprehensiveness*: The number of phenomena considered endogenously by the model. Wegener identified a total of eight phenomena, which could be simulated by using as many as other submodels: land-use distribution, transport infrastructure distribution, distribution of the residential housing supply, distribution of the non-residential real estate supply, distribution of production activities, population distribution, goods transport distribution and passenger transport distribution.
2. *The structure of the model*: Depending on whether all the sub-models are integrated under a unique theoretical principle, which unifies them, or

alternatively, whether all the sub-models are relatively autonomous and can function by using different theoretical suppositions.

3. *Theoretical base*: Depending on the fundamental theoretical base supporting the model to simulate the choices made by urban agents.

4. *Modelling techniques*: Considering the scale of the modelling (macro, meso or micro) as well as the time periods being used (e.g. 1, 5 or 10 years).

5. *Dynamics*: Whether the model finds the solution to a simulation based on equilibrium between supply and demand, and whether the model is dynamically based on the hypothesis of the existence of disequilibrium in the system.

6. *Data requirements*: Considering whether or not the model requires a greater or lesser amount of data to run the simulation.

7. *Calibration and validation*: Depending on the calibration procedures for the parameters used in the base situation and the validation of the model when compared with the data that was not used for the calibration.

8. *Operability*: Depending on whether the model is used mainly for research purposes or it has been applied to real planning situations in different study areas.

9. *Applicability*: Considering the number of questions and problems, which the model can help in answering.

Of all the criteria highlighted by Wegener, the consideration of the model's theoretical base for its classification has often been used in various studies by Foot (1981), Anas (1987) and Waddell and Ulfarsson (2004). Iacono et al. (2008), in turn, provided a classification of LUTI models developed in the literature, which was also based on the chronological development of their research and considered, therefore, the appearance of the different theoretical paradigms that have configured the field of transport and land use modelling. Foot (1981) also contributed other possible criteria for classifying the different LUTI models: whether they are deterministic or stochastic, purely static or comparative static equilibrium, the level of aggregation used and whether they are simulation models with a descriptive/explanatory nature or optimisation models with a more or less normative nature.

2.2 LUTI MODEL CLASSIFICATIONS FOUND IN THE LITERATURE

One of the first published LUTI model classifications was made by Foot (1981) who classified the operational LUTI models designed up to the 1980s into four main types considering the theoretical base on which they had been developed. These four types were:

- *Gravity models based on spatial interaction theory*: For example, the Garin–Lowry model (Garin 1966, Lowry 1964).
- *Linear regression models*: Such as the EMPIRIC model (Hill 1965).
- *Models based on optimisation techniques*: For example, the Technique for Optimum Placement of Activities into Zones (TOPAZ) model (Brotchie et al. 1980) (Chapter 7).

- *Hybrid models*: Models combining elements from the three previous types – extensions of the Garin–Lowry model such as the Echenique (Echenique et al. 1974) model or the Projective Land Use Model (PLUM) (Putman 1979).

Anas (1987) presented an alternative classification to Foot, which was also based on a model's theoretical nucleus. Anas identified the following five types of urban and regional models:

- *Monocentric models*: Models derived from urban economic theory based on the hypothesis of employment concentration at a single urban centre. The most representative case is the Alonso model and its extensions (Alonso 1964) (Chapter 4).
- *Non-economic models*: Models that have no direct derivation from economic theory, for example, the Lowry model.
- *Models based on mathematical programming*: Models based on optimisation such as the Herbert–Stevens model (Herbert and Stevens 1960a) and the TOPAZ model.
- *Models based on econometric techniques*: Models tied to urban economic theory and estimable by standard econometric techniques using statistical data. An example would be the NBER model (Ingram et al. 1972).
- Models applied to large urban or regional areas based on input–output matrices such as the MEPLAN model.

Wegener (2004) classified 20 urban models considered to be representative of the state of the art at that time. These models were: BOYCE, CUFM, DELTA, ILUTE, IMREL, IRPUD, ITLUP, KIM, LILT, MEPLAN, METROSIM, MUSSA, PECAS, POLIS, RURBAN, STASA, TLUMIP, TRANUS, TRESIS and UrbanSim. Wegener differentiated them according to the following characteristics, considering only the theoretical foundations of the models:

- *Considering the real estate market*: Eleven models simulated the land or housing market using prices endogenous to the model and a market clearing mechanism: DELTA, IMREL, KIM, MEPLAN, METROSIM, MUSSA, PECAS, RURBAN, TLUMIP, TRANUS and TRESIS. As well as simulating the real estate market using endogenous prices, three of the models, ILUTE, IRPUD and UrbanSim considered as the delays caused by pricing adjustments due to changes in supply and demand. Finally, three models were hybrids in the sense that they simulated the functioning of the real estate market combining the theory of the bid–rent price with random utility theory to stochastically simulate the prices for each location: MUSSA, RURBAN and STASA.
- *Considering equilibrium in the system*: Two models determined a general equilibrium in both the transport sector and the land use sector: KIM and METROSIM. Four models only considered equilibrium in the transport sector: ILUTE, IRPUD, ITLUP and TLUMIP. Finally, five models considered equilibrium in both the transport sector and the land use sector, but they did it separately: IMREL, MEPLAN, PECAS, TRESIS and TRANUS.

- *Considering the application of other specific theories*: The IMREL model used optimisation techniques applied to the maximisation of location surplus. Four models used economic base theory and input–output matrices to link the location of economic activities with the location of employment: MEPLAN, METROSIM, PECAS and TRANUS. Six models used cohort survival analysis to analyse the evolution of population and households: DELTA, ILUTE, IRPUD, LILT, TLUMIP and UrbanSim. One model also applied concepts obtained from time geography (Hägerstrand 1985) such as time restrictions and action spaces: IRPUD.

Waddell and Ulfarsson (2004) differentiated the following approaches to modelling used in LUTI models:

- Spatial-interaction models such as ITLUP (Putman 1983).
- Models based on spatial input–output matrices such as MEPLAN (Echenique 1994) or TRANUS (Barra 1989).
- Models based on optimisation and linear programming such as TOPAZ and POLIS (Brotchie et al. 1980).
- Microsimulation models such as MASTER (Mackett 1992), DORTMUND (Wegener 1985) or UrbanSim.
- Models based on discrete choices such as UrbanSim (Waddell 2002).
- Models based on land development rules such as CUF (Landis 1994) and WHATIF (Klosterman 1999).
- Activities models such as TRANSIMS (Barrett et al. 2002).
- Multi-agent simulation models such as SWARM (Minar et al. 1996).

In a more recent publication, Iacono et al. (2008) performed a state-of-the-art review on the LUTI models that had been applied from the 1960s to the present day. The authors made their classification based on the chronological evolution of the different theoretical paradigms that used to model the interaction between land use and transport. They were able to differentiate

- *Models based on spatial interaction theory*: Lowry model, the LILT model (Mackett 1991), the IRPUD model (Wegener 1982), the ITLUP model and its follow-up model METROPILUS (Putman and Chan 2001).
- Models based on random utility theory, differentiating into two large subtypes they found: regional economic models and real estate market models. The MEPLAN and TRANUS models belonged to the first group, whereas the CATLAS model could be found in the second group (Anas 1987) and their later version, the METROSIM model as well as the MUSSA (Martínez 1997) and UrbanSim models.
- The disaggregated micro-simulation models were, in turn, subdivided into activity-based models such as TRANSIMS, models based on individual agents such as ILUMASS or ILUTE (Salvini and Miller 2005) and cellular automata models such as MALUT and LUCI2 (John 2005).

To summarise, it can be seen that over the years the research developed by different authors and research groups have found different categories for classifying LUTI models. However, there is a lot of coincidence between the classifications and the authors considered here have generally chosen to use the theoretical foundations of the models as the fundamental principle behind their classifications.

2.3   A PROPOSED CLASSIFICATION

A model typology founded on the previously described work is proposed in the following them as a function of their basic theoretical nucleus and the generation they belonged to when developed. The three generations of models and five basic types have been differentiated (Figure 2.1).

1. *First-generation models*: These are the models that appeared earliest in the 1960s and 1970s. They can be divided further into three basic types according to the theoretical foundation that they used to simulate how the urban system functions.
   a. *Spatial interaction or gravity models*: Models based on spatial interaction theory or on the generalisation based on Wilson's statistical mechanics (1970). The most classic example is Lowry's interaction model, which was also based on economic base theory (Andrews 1953) to simulate the location of population and economic activities.

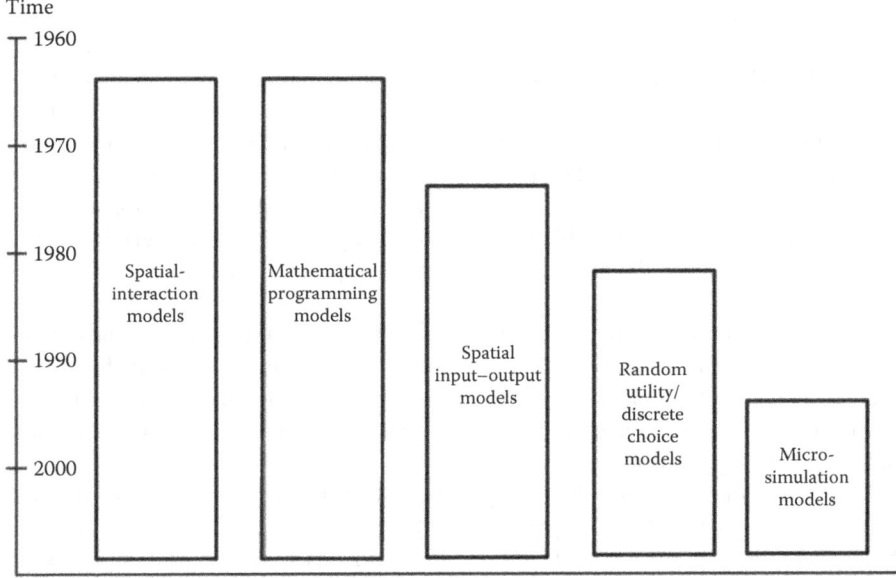

FIGURE 2.1   Chronological development of LUTI models. (Based on Iacono, M. et al., *J. Plan. Lit.*, 22, 323–340, 2008.)

    b. *Mathematical programming models*: Models based on optimisation techniques. This kind of model is founded on a simulation of agent behaviour through the minimisation or maximisation of a certain objective. The classic model of this type was developed by Herbert and Stevens (1960b) and coherently simulated the workings of the residential location market, based on Alonso's theory, through the maximisation of aggregated rents. Another example of a model of this type is TOPAZ, which determined the locations of activities as a function of the minimisation of transport and urban development costs (Brotchie et al. 1980).

    c. *Models based on INPUT/OUTPUT matrices*: This type of model simulates the urban or regional economy; thanks to the technique of input–output matrices developed from the work of Leontief (1966). MEPLAN is a well-known example of this type of model (Echenique 1994, 2011).

2. *Second-generation models*: These models appeared during the 1980s and 1990s and are based on random utility theory developed from the work of McFadden (1974). The real estate market simulation models using random utility theory, which appeared from the work of Anas (1982), can also be differentiated within this generic type. The Santiago Land Use Model (MUSSA) developed by Martínez (1997) is another example of a second-generation model.

3. *Third-generation models*: These models appeared more recently around the second half of the 1990s. They are highly disaggregated models, which some authors have called micro-simulation models (Iacono et al. 2008). They have a dynamic character; in other words, they do not acquire complete market equilibrium as a solution for the simulations. On account of its wide application, a highlight among the well-known models of this kind is UrbanSim developed by Waddell and collaborators from the University of Washington (Waddell et al. 2007). Models that developed following the alternative theoretical paradigms can also be included in this category, such as activity models based on the description of users' daily journey patterns (Castiglione et al. 2015) and the cellular automata models that simulate land use as a group of cells, which experience change depending on certain rules of behaviour and the state of the neighbouring cells (Liu 2008).

Random utility theory has become the most used paradigm for modelling the location choices of different urban agents. This theory has largely substituted the spatial-interaction location models, which offered a smaller behavioural base, although as various researches have shown, both approaches can produce equivalent results under specific suppositions (Anas 1983).

    In the field of simulating the transport sub-system, the LUTI models have been criticised on occasions for incorporating methods, which are somewhat distant from the state of the art. Many LUTI models are still based on the classic four-stage sequential approach, which has led many authors to recommend the use of more modern models whether they are endogenous or exogenous to the rest of the LUTI simulator (Wegener 2004). Table 2.1 shows a summary of the main characteristics and references of 19 LUTI models classified according to their theoretical structure.

TABLE 2.1

Summary of LUTI Models

Name	Characteristics	References
	Models Based on Spatial Interaction Theory	
LOWRY	First operational LUTI model	Lowry (1964)
GARIN–LOWRY	Matricial representation of the Lowry model	Garin (1966)
ITLUP/METROPILUS	Integrated transportation and land use package. Improved spatial interaction model considered to be the first operational software package. The METROPILUS model incorporated an improved interface based on a geographical information system	Putman (1983), Putman and Chan (2001)
LILT	Leeds integrated land use. Combination of a Lowry-type model with a four-stage transport model. LILT also included a car ownership choice model	Mackett (1983)
IRPUD	Seven interrelated models, including supply and demand of economic activities and population located using gravity models	Wegener (1982)
	Models Based on Mathematical Programming	
HERBERT–STEVENS	First model that made Alonso's land use theory operational for the case of residential location	Herbert and Stevens (1960a)
TOPAZ	Technique for optimal placement of activities into zones. Location model based on the maximisation of net gain coming from spatial interactions and land use	Brotchie et al. (1980)
POLIS	Projective optimisation land-use information system. Residential choice model using non-linear programming incorporating three functions of surplus according to work place location, journey to work mode and shopping behaviour	Caindec and Prastacos (1995)
IMREL	Interrelationship between a residential location sub-model, a work place location sub-model and a transport sub-model	Anderstig and Mattsson (1998)

(Continued)

(*Continued*)

TABLE 2.1 (*Continued*)
Summary of LUTI Models

Name	Characteristics	References
Models Based on INPUT/OUTPUT Matrices		
MEPLAN	Model in which the activities are located using discrete choice models starting from an input–output matrix	Echenique et al. (1990)
TRANUS	Similar to MEPLAN, TRANUS also starts from an input–output matrix to simulate the behaviour of a regional economy, which later becomes spatial using discrete choice models	Barra (1989, 1994)
Models Based on Random Utility Theory		
CATLAS	Chicago area transportation-Land use analysis system. Model designed to evaluate the impact changes made to transport that have on land use and modal choice. Behaviour models are based on discrete choice logit models.	Anas (1987)
METROSIM	Improved version of CATLAS incorporating residential and commercial space market models	Anas (1994)
MUSSA	The Santiago land use model. This model combines random utility theory with the Alonso bid–rent theory to simulate the housing market and locate the demand for land	Martinez (1996)

TABLE 2.1 (*Continued*)
Summary of LUTI Models

Name	Characteristics	References
	Models Based on Microsimulation	
ILUMASS	Integrated land use modelling and transportation system simulation. Integrates a dynamic traffic simulator with land use and environmental models to simulate the impact of transport on the environment	Strauch et al. (2005)
ILUTE	Integrates four sub-models: land use, location choice, car ownership and journey-activity patterns. It uses a plurality of modelling techniques, including discrete choice, rule-based models, transition models and others	Salvini and Miller (2005)
UrbanSim	An open-coded model with modular architecture. Allows simulation of real estate market at a zonal level and plot level. UrbanSim has sub-models for residential location, economic activity location, urban development location and real estate prices	Waddell et al. (2003)
TRANSIMS	Transportation analysis and simulation system. Activity-based model. Includes the following sub-models: a population model, an activity generation model, a route planning model and a traffic micro-simulation model	Barrett et al. (2002)
LUCI2	Land use in Central Indiana. Model based on cellular automata to simulate the evolution of land use. It is an extension of the LUCI model and is able to simulate the location of population and employment separately	John (2005)

REFERENCES

Alonso, W. 1964. *Location and Land Use: Toward a General Theory of Land Rent, Publications of the Joint Center for Urban Studies of the Massachusetts Institute of Technology and Harvard University*. Cambridge, UK: Harvard University Press.

Anas, A. 1982. *Residential Location Markets and Urban Transportation: Economic Theory, Econometrics, and Policy Analysis with Discrete Choice Models, Studies in Urban Economics*. New York: Academic Press.

Anas, A. 1983. Discrete choice theory, information theory and the multinomial logit and gravity models. *Transportation Research Part B: Methodological* 17 (1):13–23. doi:10.1016/0191-2615(83)90023-1.

Anas, A. 1987. *Modeling in Urban and Regional Economics, Fundamentals of Pure and Applied Economics*. New York: Harwood Academic Publishers.

Anas, A. 1994. *METROSIM: A Unified Economic Model of Transportation and Land-Use*. Williamsville, NY: Alex Anas & Associates.

Anderstig, C. and Mattsson, L.-G. 1998. Modelling land-use and transport interaction: Policy analyses using the IMREL model. In *Network Infrastructure and the Urban Environment*, Tschangho, J. K. and Lars-Göran, M. (Eds.), pp. 308–328. Berlin, Germany: Springer.

Andrews, R. B. 1953. Mechanics of the urban economic base: Historical development of the base concept. *Land Economics* 29 (2):161–167.

Barra, T. 1989. *Integrated Land Use and Transport Modelling: Decision Chains and Hierarchies, Cambridge Urban and Architectural Studies*. Cambridge, UK: Cambridge University Press.

Barra, T. 1994. From theory to practice: The experience in Venezuela. *Environment and Planning B: Planning and Design* 21:611–617.

Barrett, C. L., Beckman, R. J., Berkbigler, K. P., Bisset, K. R., Bush, B. W., Campbell, K., Eubank, S. et al. 2002. TRansportation ANalysis SIMulation system (TRANSIMS). Portland Study Reports, December 10, 2002.

Brotchie, J. F., Dickey, J. W. and Sharpe, R. 1980. *TOPAZ: General Planning Technique and its Applications at the Regional, Urban, and Facility Planning Levels, Lecture Notes in Economics and Mathematical Systems 180*. Berlin, Germany: Springer-Verlag.

Caindec, E. K. and Prastacos, P. 1995. A description of POLIS. The projective optimization land use information system. Working Paper 95-1, Oakland, CA: Association of Bay Area Governments.

Castiglione, J., Bradley, M. and Gliebe, J. 2015. *Activity-Based Travel Demand Models: A Primer*. SHRP 2 Report S2-C46-RR-1. Transportation Research Board.

Echenique, M., Feo, A., Herrera, R. and Riquezes, J. 1974. A disaggregated model of urban spatial structure: Theoretical framework. *Environment and Planning A* 6 (1):33–63.

Echenique, M. H. 1994. Urban and regional studies at the Martin centre: Its origins, its present, its future. *Environment and Planning B: Planning and Design* 21:517–533.

Echenique, M. H. 2011. Land use/transport models and economic assessment. *Research in Transportation Economics* 31:45–54.

Echenique, M. H., Flowerdew, A. D. J., Hunt, J. D., Mayo, T. R., Skidmore, I. J. and Simmonds, D. C. 1990. The MEPLAN models of bilbao, leeds and dortmund. *Transport Reviews* 10 (4):309–322.

Foot, D. H. S. 1981. *Operational Urban Models: An Introduction*. London, UK: Methuen.

Garin, R. A. 1966. A matrix formulation of the lowry model for intrametropolitan activity allocation. *Journal of the American Institute of Planners* 32 (6):361–364. doi:10.1080/01944366608978511.

Hägerstrand, T. 1985. Time-geography: Focus on the corporeality of man, society, and environment. In *The Science and Praxis of Complexity*, Aida, S. (Ed.), pp. 193–216. Tokyo Japan: United Nations University Press.

Herbert, J. D. and Stevens, B. H. 1960a. A model for the distribution of residential activity in urban areas. *Journal of Regional Science* 2 (2):21–36. doi:10.1111/j.1467-9787.1960. tb00838.x.

Herbert, J. D. and Stevens, B. H. 1960b. A model for the distribution of residential activity in urban areas. *Journal of Regional Science* 2 (2):21–39.

Hill, D. M. 1965. A growth allocation model for the Boston region. *Journal of the American Institute of Planners* 31 (2):111–120.

Iacono, M., Levinson, D. and El-Geneidy, A. 2008. Models of transportation and land use change: A guide to the territory. *Journal of Planning Literature* 22 (4):323–340.

Ingram, G. K., Kain, J. F. and Ginn, J. R. 1972. Front matter, the Detroit prototype of the NBER urban simulation model. In *The Detroit Prototype of the NBER Urban Simulation Model*, Ingram, G. K., John, F. K. and Royce, G. J. (Eds.), pp. 103–128. Cambridge, MA: NBER Books.

John, R. O. 2005. Accessibility in the Luci2 urban simulation model and the importance of accessibility for urban development. In *Access to Destinations*, Levinson, D.M. and Krizek, K.J. (Eds.), pp. 297–324. Amsterdam, the Netherlands: Elsevier.

Klosterman, R. E. 1999. The what if? Collaborative planning support system. *Environment and planning B: Planning and design* 26 (3):393–408.

Landis, J. D. 1994. The California urban futures model: A new generation of metropolitan simulation models. *Environment and planning B: planning and design* 21 (4):399–420.

Leontief, W. 1966. *Input-Output Economics*. New York: Oxford University Press.

Liu, Y. 2008. *Modelling Urban Development with Geographical Information Systems and Cellular Automata*. Boca Raton, FL: CRC Press.

Lowry, I. S. 1964. *A Model of Metropolis, Memorandum*. Santa Monica, CA: Rand Corporation.

Mackett, R. 1992. *Micro Simulation Modelling of Travel and Locational Processes: Testing and Further Development*. Report to the Transport and Road Research Laboratory, March. London: Transport Studies Group, University College London.

Mackett, R. L. 1983. *Leeds Integrated Land-Use Transport Model (LILT)*. Report to the Transport and Road Research Laboratory. Crowthorne, UK: Special Research Branch, Safety and Transportation Department.

Mackett, R. L. 1991. LILT and MEPLAN: A comparative analysis of land-use and transport policies for Leeds. *Transport Reviews* 11 (2):131–154.

Martinez, F. 1996. MUSSA: Land use model for Santiago city. *Transportation Research Record* 1552:126–134.

Martínez, F. 1997. MUSSA: Land use model for Santiago city. *Transportation Research Record* 1552. doi:10.3141/1552-18.

McFadden, D. 1974. Conditional logit analysis of qualitative choice behaviour. In *Frontiers in Econometrics*, Zarembka, P. (Ed.), pp. 105–142. New York: Academic Press.

Minar, N., Burkhart, R., Langton, C. and Manor, A. 1996. The swarm simulation system: A toolkit for building multi-agent simulations, Working Paper, 96-06-042, Santa Fe NM: Santa Fe Institute.

Putman, S. H. 1979. *Urban Residential Location Models, Studies in Applied Regional Science*, Vol. 13. Boston, MA: Martinus Nijhoff Publishers.

Putman, S. H. 1983. *Integrated Urban Models*. 2 vols, Vol. 1: *Research in planning and design*. London, UK: Pion.

Putman, S. H. and Chan, S.-L. 2001. The METROPILUS planning support system: Urban models and GIS. In *Planning Support Systems: Integrating Geographic Information Systems, Models, and Visualization Tools*, Brail, R.K. and Klosterman, R.E. (Eds.), pp. 99–128. Redlands, CA: ESRI Press.

Salvini, P. and Miller, E. J. 2005. ILUTE: An operational prototype of a comprehensive micro-simulation model of urban systems. *Networks and Spatial Economics* 5 (2):217–234.

Strauch, D., Moeckel, R., Wegener, M., Gräfe, J., Mühlhans, H., Rindsfüser, G. and Beckmann, K.-J. 2005. Linking transport and land use planning: The microscopic dynamic simulation model ILUMASS. In *Geodynamics*, Atkinson, P.M., Foody, G.M., Darby, S.E. and Wu, F. (Eds.), pp. 295–311. Boca Raton, FL: CRC Press.

Waddell, P. 2002. UrbanSim: Modeling urban development for land use, transportation, and environmental planning. *Journal of the American Planning Association* 68 (3):297–343.

Waddell, P., Borning, A., Noth, M., Freier, N., Becke, M. and Ulfarsson, G. 2003. Microsimulation of urban development and location choices: Design and implementation of urbansim. *Networks and Spatial Economics* 3 (1):43–67.

Waddell, P. and Ulfarsson, G. F. 2004. Introduction to urban simulation: Design and development of operational models. In *Handbook in Transport, Volume 5: Transport Geography and Spatial Systems*, Stopher, B. and Kingsley, H. (Eds.), pp. 203–236. Oxford, UK: Pergamon Press.

Waddell, P., Ulfarsson, G. F., Franklin, J. P. and Lobb, J. 2007. Incorporating land use in metropolitan transportation planning. *Transportation Research Part A: Policy and Practice* 41 (5):382–410.

Wegener, M. 1982. Modeling urban decline: A multilevel economic-demographic model for the Dortmund region. *International Regional Science Review* 7 (2):217–241.

Wegener, M. 1985. The Dortmund housing market model: A Monte Carlo simulation of a regional housing market. In *Microeconomic Models of Housing Markets*, Konrad, S. (Ed.), pp. 144–191. Berlin, Germany: Springer.

Wegener, M. 2004. Overview of land-use transport models. In *Transport Geography and Spatial Systems*, David, A. H. and Kenneth, B. (Eds.), pp. 127–146. Kidlington, UK: Pergamon/Elsevier Science.

Wilson, A. G. 1970. *Entropy in Urban and Regional Modelling, Monographs in Spatial and Environmental Systems Analysis 1*. London, UK: Pion.

Section II

The Theoretical Basis of Land Use–Transport Interaction Models

This section provides a review of the theoretical basis behind the land use–transport interaction (LUTI) model simulation. Chapter 3 concentrates on accessibility as the primary link between the transport system and the land use system. Chapter 4 provides a brief review of the microeconomic theory underpinning the theoretical relationship between transport and land use, whereas Chapters 5–7 address three paradigms that have been widely applied in the field of operational LUTI models: (1) spatial interaction theory, (2) random utility theory and (3) the optimisation approach.

3 The Nexus between Transport and Land Use: Accessibility

Hugo González, Rubén Cordera and Ángel Ibeas

CONTENTS

3.1 THE RELEVANCE OF ACCESSIBILITY

Making a journey between two zones on a transport network depends on the following three fundamental elements (El-Geneidy and Levinson 2006):

1. The potential of the origin zone to generate the journey. This variable will further depend on the number of inhabitants in the zone, the population's level of motorisation and their standard of living along with other factors.
2. The capacity of the destination zone to attract a trip. The attraction depends on the population, the number of commercial premises, the number of jobs and other facilities present in the destination zone.
3. The difficulty in travelling between the origin and destination zones. In order for this to affect the journey, the effect of this cost has to be lower than the destination's power of attraction.

Depending on the scale of the study, the zones considered to be the producers and attractors of trips may correspond to more or less large areas within a city or even entire population nuclei if the research is regional.

Better standards of infrastructure and transport services will modify the journey cost and enable communication between zones, in other words, improve a component

25

of accessibility. The cost of a journey is a basic variable as shorter journey times and less money required for travelling means that more places become reachable within a determined budget, thereby increasing accessibility.

The destinations and the available activities they offer are also important: the greater the number of possible destinations, the greater the variety of available opportunities which, again, will improve accessibility. The modal choice for making the journey is equally crucial: a wider range of the possible transport modes for reaching the destination means greater choice for the user, resulting in an increased accessibility.

The concept of accessibility has formed a part of transport and land use planning for more than 50 years. One of the first attempts at applying it in the planning field was made by Hansen (1959), who defined accessibility as a measure of the intensity of the possibility of an interaction. Other similar definitions also differentiate two components of accessibility: (1) an element of transport (hindrance or cost) and (2) an activity that motivates or attracts the journey (opportunity or profit) (Burns 1980, Koenig 1980). More recently, authors like Batty (2009) distinguished between the relative accessibility of one place to another from the calculated accessibility from one place to all the other places. In the latter case, the resulting accessibility index provides a synthetic measurement of the ability of reaching a certain type of opportunity from a place of origin.

Thus, accessibility may be defined as a measurement of the capacity to communicate between human activities or settlements using a determined transport system. The usual measurement units are distance, time, journey cost and the number of opportunities or adimensional variables resulting from their interrelationships. By calculating the accessibility levels that are provided by a transport network, the planner or decision-maker has a tool that allows them to

- Identify the places with least accessibility and therefore fewer possibilities of mobility and access to opportunities.
- Compare alternative transport plans to evaluate if they further aim the increasing accessibility and promoting territorial equality between different places.
- Evaluate the impact of different planning alternatives on accessibility.
- Show graphic representations of the results of different planning alternatives to improve their comprehension, particularly for non-technical people.

For these reasons, accessibility indicators have been used in the evaluation of land use and transport plans and policies for decades. One of the first examples of the use of an accessibility indicator can be found in the work of Hansen (1959). Hansen analysed the location and residential development of the metropolitan area of Washington DC, with reference to only two variables: (1) the amount of available land in each zone and (2) the accessibility to employment. A more recent example has been presented by Waddell et al. (2007) for the land use–transport interaction (LUTI) model UrbanSim. This model uses accessibility as the nexus for the relationship between land use and transport. In UrbanSim, the accessibility indicator is a function of the activities of households and the disutility of reaching them using the transport system. The indicator is then introduced as one more variable for locating the population and the activities within the studied urban system in such a way that the more accessible zones are the most attractive. This methodology has been applied in a similar way using other LUTI models.

Some authors have addressed the question of whether an increased accessibility continues to be relevant in conditioning the location of population and economic activities in areas, which are already well developed with high levels of mobility and availability of opportunities (Giuliano 1995). The more commonly accepted conclusion is that accessibility continues to be a significant factor (Cervero and Landis 1995). Nevertheless, it should never be considered as the sole factor as others such as the prevailing local environmental conditions (noise, air pollution, presence of green zones and others), the availability of land and the local regulations are also very relevant when simulating the location of urban agents (Badoe and Miller 2000).

3.2   FACTORS THAT INFLUENCE ACCESSIBILITY

Measures of accessibility are indicators reflecting the impact that land use distribution and the characteristics of the transport system have on the users. This means that both the terms, land use and transport, ought to be related because they give people the opportunity of participating in activities occurring in different places. The four basic components can be identified when measuring accessibility (Geurs and van Wee 2004):

- The *land use* component refers to the spatial distribution of the opportunities offered at each destination, in terms of quantity and quality: jobs, retail areas, health centres and so on; the demand generated by these opportunities in the place of origin (where the population lives); and the resulting confrontation between both supply and demand.
- The *transport* component describes the transport system as well as the cost against an individual covering a distance between an origin and a destination using a specific mode of transport and for each one includes the amount of time (journey time, waiting time, stoppage time), the monetary costs (fixed and variable) and the quality of the mode (reliability, comfort levels, risk of accidents, etc.). The supply provided by the infrastructure includes its location and its characteristics (maximum permitted speed, number of lanes, type of road, public transport timetables, cost of journey, etc.).
- The *time* component reflects the time limitations, the availability of opportunities at different times of the day (rush hour, off-peak) and the users' availability to participate in certain activities (work, leisure, etc.).
- The *individual* component refers to the needs (depending on age, income, education, etc.), the abilities (depending on the physical condition of the person, availability of certain modes of transport, etc.) and the opportunities (also depending on income levels, journey cost, etc.) of the individuals. These characteristics influence the levels of access a person has in the

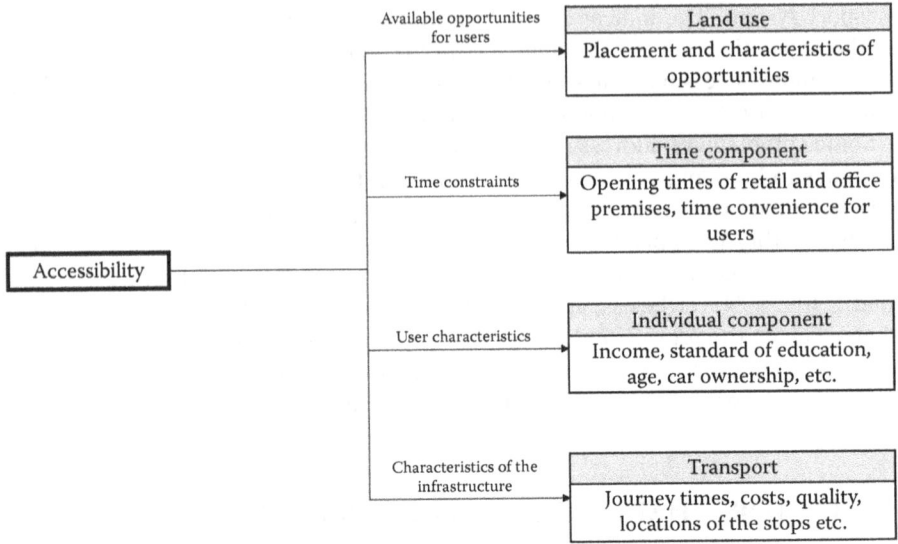

FIGURE 3.1   Relationship between accessibility and its components.

different available transport modes (the possibility of using or sharing a car; to have the ability or the education required to be qualified for a job opportunity close to their zone of residence) and they can also have a significant influence on overall accessibility.

These four components provide different characteristics when evaluating accessibility (Figure 3.1).

3.3   TYPES OF INDICATORS AVAILABLE FOR MEASURING ACCESSIBILITY

Before analysing the most frequently used accessibility indicators, it would be beneficial to know the scale and coverage of the study as comparisons can only really be drawn when they coincide. Three levels of accessibility can be distinguished:

- *Relative accessibility* (a_{ij}): Measures the quality of the connection between the two points located in the same territory.
- *Integral accessibility* (A_i): Measures the degree of interconnection of node i with all the other nodes j within the same study area as

$$A_i = \sum_j a_{ij} \tag{3.1}$$

- *Overall accessibility* (A): It is the sum of the integral accessibilities of all the study zone nodes, or

$$A = \sum_i A_i \qquad (3.2)$$

Relative and integral accessibilities serve to establish comparisons between nodes in the study area and draw conclusions, whereas a zone's overall accessibility results cannot be compared with those of another as they depend on the number of nodes in each area as well as their relative position and the intervening variables. However, they can be applied for making comparisons between different transport and land use alternatives or policies within the same zone.

It is a standard practice to use maps to represent accessibility in each zone. Isoaccessibility curves can also be used to unify network nodes with the same levels of accessibility. Depending on which cost variable is being used, distance, journey time or monetary cost, these curves are known as isodistance, isochrone (Figure 3.2) or isocost curves.

The kinds of accessibility indicators that have been used to date for academic research and field application can be divided into various groups. Handy and Niemeier (1997) classified the indicators into three main types: (1) accumulated opportunities, (2) gravity analogy-based accessibility and (3) accessibility based on random utility theory. The classification presented by Geurs and van Wee (2004) differentiated four types of accessibility: (1) infrastructure-based, (2) location-based, (3) person-based – depending on their capacity to participate over a determined time period, and finally, (4) based on random utility theory. In a more recent classification,

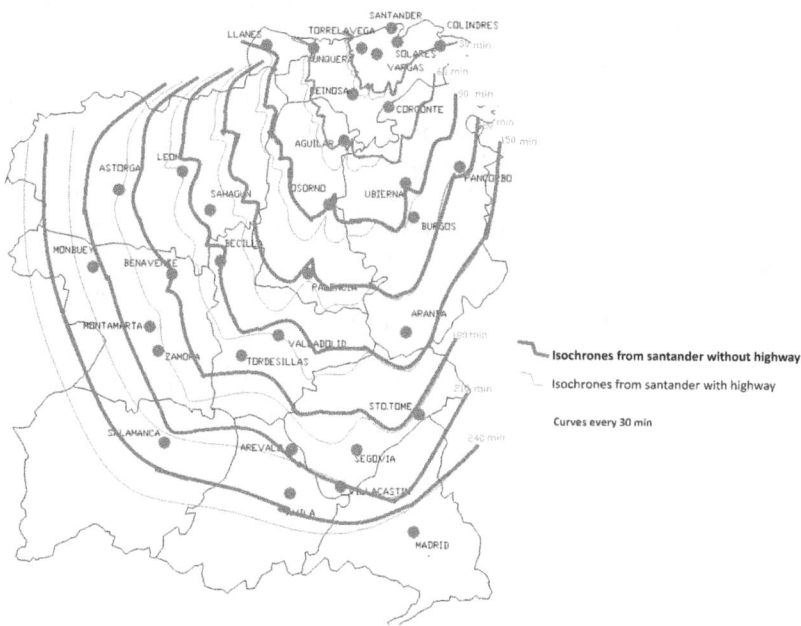

FIGURE 3.2 Example of isochrone curves.

Batty (2009) distinguishes three types of accessibility: the first kind considers the opportunities offered by a destination and the cost of reaching it using a transport network or not (this type of indicator is similar to the accumulated opportunities or gravity indicators mentioned previously); a second type is based on the infrastructure, where it is assumed that the opportunities are equal in all the destinations and only the cost from using the existing network is considered; and finally, the third and most recent kind that splits space into smaller portions (e.g. links on a network) to find indicators about their integration and connectivity. This latter kind of accessibility is derived from the work of Hillier and Hanson (1989) and has hardly been used at all with LUTI models, mainly being applied in town planning and architecture research (Hillier 2007).

If these different types of accessibility indicators are analysed according to their authors, it can be seen that in many cases the indicators measure similar phenomena, the distinction being that each author endows them with a different name. For example, accessibility as accumulated opportunities refers to the opportunities that are available within a certain range of time or distance, whereas in the classification by Geurs and van Wee (2004) it would be found among the location-based indicators.

The classification chosen in this chapter is an extension of Handy and Niemeier (1997), to which a fourth indicator has been added, simpler than the others, which analyses accessibility as a function of infrastructure criteria (see the following section).

3.3.1 Accessibility Indicators Based on the Transport System

Some of the more basic accessibility indicators, which refer only to the aspects of infrastructure such as layout, speed and running time from one zone to another, will be explained as follows:

1. The most simple accessibility indicator available in the literature is that of *presence/absence*. This indicator consists of dividing the study zone into various sub-zones, either by administrative territorial units or by superimposing a regular polygonal mesh over the study area. Once the study area is divided up, the simplest version of this indicator assigns the value of 1 or 0 to each sub-zone, depending on whether there is or there is not a form of transport present. This indicator can be perfected by adding additional values depending on the importance of the infrastructure that communicates the two points.

2. Another basic indicator is the *density index*, the commonest formulation of which is

$$D = \frac{\text{Network Kilometers}}{\text{Surface of the study area (Km}^2)} \tag{3.3}$$

This accessibility indicator reflects the density of the communication infrastructure within a zone. There are many variants, depending on the weighting placed on the routes or the importance of the infrastructure, rather than simply evaluating the network length.

3. The *route factor* is an indicator that measures the quality of the infrastructure comparing its layout with how much it mirrors a straight line, representing the quickest route between the two points:

$$r_{ij} = \frac{d_{ij}}{dg_{ij}}$$ (3.4)

where:
r_{ij} is the route factor between points i and j
d_{ij} is the minimum distance using the transport network between i and j
dg_{ij} is the physical distance in a straight line from i to j

The integral route factor can be determined as the interconnection of node i with all the other nodes present in the zone, where n is the total number of nodes (zones):

$$R_i = \frac{1}{n-1} \sum^{n} \frac{d_{ij}}{dg_{ij}}$$ (3.5)

It could be said that route factor values greater than 1.5 indicate low levels of accessibility. This is very common in mountainous areas where the terrain means there are a lot of curves along the transport infrastructure.

Finally, a variety of the route factor is the *route–velocity index* in which instead of making a comparison with the distances, it is done with travel times (Figure 3.3):

$$Itv_i = \frac{1}{n} \frac{\sum_j t_{ij}}{\sum_j t_{ij}^0}$$ (3.6)

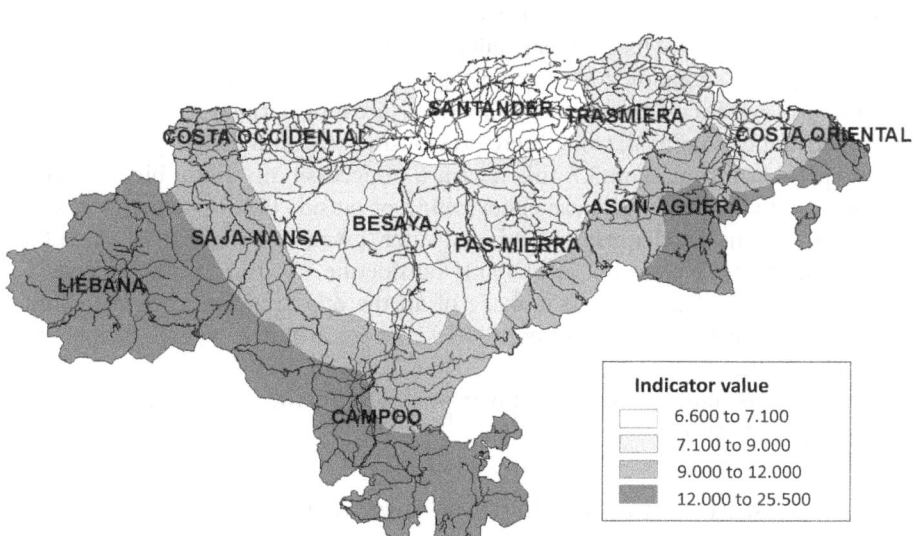

Indicator value	
	6.600 to 7.100
	7.100 to 9.000
	9.000 to 12.000
	12.000 to 25.500

FIGURE 3.3 Application example of the route–speed index.

where:

t_{ij} is the minimum travel time from i to j using the network
t^o_{ij} is the fictitious time taken to cover the distance ij at the average speed limit
n is the number of nodes in each zone.

3.3.2 Accessibility Indicators Based on Accumulated Opportunities

This group of accessibility indicators evaluates the reachable opportunities from an origin within a determined cost threshold, using the transport network or not.

This type of indicator begins by establishing a mode of transport, a cost interval and a number of opportunities that can be reached using the given mode within the present cost interval. For example, the number of reachable jobs within an interval of 15 min using a private motorised vehicle can be identified. The formula used is as follows:

$$A_i = \sum_{j=1}^{J} (B_j \cdot C_j) \qquad (3.7)$$

where:

B_j is the number of opportunities (goods, activities and population) offered in zone j
$C_j = 1$ if $C_j \le C_k$; $C_j = 0$ if $C_j > C_k$, where C_k is the cost measurement corresponding to the isochrone (or isocost) of the value k (10, 20, 30… min, distances or any other measure of cost)

The methodology followed by these types of indicators can be used to capture the influence that the land use and transport system have on accessibility, given that they have an effect on the reachable opportunities from the origin. For example, the construction of a new shopping centre in a determined zone represents new opportunities; or improvements made to existing infrastructure reduce running time and make new opportunities available, which were previously not within the range.

As expressed in Equation 3.7, this measure of accessibility has a binary nature meaning that the opportunities are either within or outside the set range. Such simplicity is the main disadvantage of this indicator, given that it does not value the opportunities that are close but just slightly outside the isocost curve being considered. For example, if an analysis is made of all the opportunities available within 20 min walking distance, all the opportunities that are 21 min away will be ignored, risking bias in the indicator's results.

To correct this setback, there are gravity style accessibility indicators, which will be explained in Section 3.3.3.

3.3.3 Accessibility Indicators Based on the Gravity Analogy

Gravity indicators follow a formulation to calculate accessibility, which places a weighting on all the opportunities and thereby can include all the possible destinations in the study area. The weighting of the opportunities present in the destination is established as a function of the cost required to reach them. At first, according to classic accessibility studies, this function was analogous with Newton's second law, the law of universal gravity, and the weight of each destination was inversely proportional to the square of the distance needed to reach it.

This particular functional form was later made more flexible by allowing the use of any functional form, by which the opportunities are reduced as journey cost increases:

$$A_i = \sum_{j=1}^{J} \left[B_j \cdot f(C_{ij}) \right] \qquad (3.8)$$

where:

A_i is the accessibility in zone i with respect to all opportunities B present in the zones of destination j

C_{ij} is the cost that could be time, distance or monetary cost from zone i to zone j

$f(C_{ij})$ is the cost function

Different cost functions have been tested by several studies such as the potential based on the physical analogy aforementioned, the Gaussian functions or the logistical functions. However, the negative exponential function is the one that is more frequently used (Handy and Niemeier 1997) (Figure 3.4 as an example).

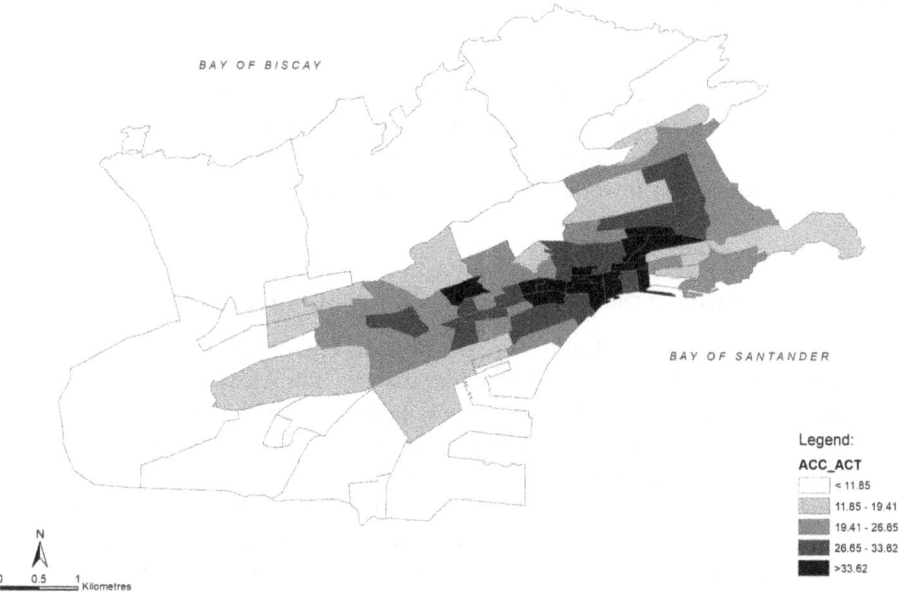

FIGURE 3.4 Example showing the application of a gravity indicator for accessibility to employment.

A variant of this widely used gravity accessibility indicator due to the simplicity of its formulation is *demographic accessibility*, taking the form

$$A_i = P_i \sum_{j=1}^{n} \frac{P_j}{t_{ij}}$$

(3.9)

where:
  A_i is the accessibility of zone i
  P_i is the population of origin zone i
  P_j is the population of destination zone j
  t_{ij} is the journey time between i and j

Normal practice in the applied field is to analyse the repercussions for economic development in a study area due to changes in the level of accessibility. One of the simplest approaches consists of correlating the demographic accessibility with the income of each sub-zone using a potential or exponential type of model.

An indicator of the income level of the study area could be found using an estimation or an indirect value such as the budget of the organisation administering the territory. A formulation of this model could be

$$R = a(A_i)^b$$

(3.10)

where:
  R is the income level
  a and b are parameters to be estimated

Equation 3.10 can be made linear by taking logarithms to both sides of the expression. Once a and b are adjusted and knowing the increase in accessibility from the difference between the scenario and the base situation, the expected increase in wealth in the study area due to changes in accessibility can be calculated, for example, due to the introduction of a new infrastructure service (Ibeas et al. 2002).

3.3.4 ACCESSIBILITY INDICATORS BASED ON INDIVIDUAL UTILITY

Researchers have also delved into other possible disaggregated measures for accessibility, such as those based on random utility theory (Ben-Akiva and Lerman 1985). According to random utility theory, the possibility that an individual chooses one particular alternative depends on the utility of that particular alternative compared to the utility of all the other available alternatives.

However, different individuals do not necessarily have to have the same alternatives available to them nor have the same homogenous preferences faced with the same characteristics. This means that faced with the same choice situation between the two alternatives, two different people can make opposite choices due to their different preferences or simply due to the random component present in each choice, which could not be captured by the model.

If it is assumed that an individual assigns a utility to each possible destination from within a group, and they later choose the alternative, which maximises their utility, then accessibility could be defined as the expected maximum utility from the choice set, also known as *logsum* (Ben-Akiva and Lerman 1985, McFadden 1980).

Accessibility, A_n, for an individual n, within the choice set C, would be

$$A_n = \ln\left[\sum_{\forall c \in C_n} \exp(V_{n(C)})\right] \tag{3.11}$$

where $V_{n(C)}$ is the systematic utility of each one of the alternatives present in the choice set C.

The specified utility function includes variables representing the attributes of each possible choice such as the attraction of destinations in terms of opportunities (jobs, population and others) as well as the cost that needs to be overcome to reach said destinations. Socioeconomic variables referring to the individual can also be included to capture changes in specific utility and even include variations in the preferences of certain attributes for the determined socioeconomic groups.

3.4 CONCLUSION

Many accessibility indicators can be found in the literature, ranging from those based only on the route or category of the infrastructure, to others such as accumulated opportunities that analyse the destinations reachable from an origin within a certain threshold. The indicators based on the gravity analogy remove the limitation of having to fix a limit, weighting the opportunities by the cost of reaching them and thereby being able to include all the possible destinations in the study area.

These three types of indicators do not consider the characteristics and constraints of the users when they choose from the available opportunities. Therefore, further disaggregated indicators were developed based, for example, on utility, in which two people located in the same area can have a different accessibility level according to their socio-economic characteristics.

The end goal of these indicators is to be able to evaluate the accessibility characteristics of an area and the changes that could occur in response to the application of measures related to land use and transport. This makes them useful tools for evaluating different policies and projects. The choice of indicator will depend on the level of detail that is required for the study and on the resources available for data collection.

REFERENCES

Badoe, D. A. and Miller, E. J. 2000. Transportation-land-use interaction: Empirical findings in North America, and their implications for modeling. *Transportation Research Part D: Transport and Environment* 5 (4):235–263.

Batty, M. 2009. Accessibility: In search of a unified theory. *Environment and Planning B: Planning and Design* 36 (2):191–194.

Ben-Akiva, M. E. and Lerman, S. R. 1985. *Discrete Choice Analysis: Theory and Application to Travel Demand*, MIT Press series in transportation studies 9. Cambridge, MA: MIT Press.

Burns, L. D. 1980. *Transportation, Temporal, and Spatial Components of Accessibility.* Lexington, MA: Lexington Books.

Cervero, R. and Landis, J. 1995. The transportation-land use connection still matters. *Access* 7:2–10.

El-Geneidy, A. M. and Levinson, D. M. 2006. *Access to Destinations: Development of Accessibility Measures.* New York: Citeseer Publisher.

Geurs, K. T. and van Wee, B. 2004. Accessibility evaluation of land-use and transport strategies: Review and research directions. *Journal of Transport Geography* 12 (2):127–140.

Giuliano, G. 1995. The weakening transportation-land use connection. *Access* 6:3–11.

Handy, S. L. and Niemeier, D. A. 1997. Measuring accessibility: An exploration of issues and alternatives. *Environment and Planning A* 29 (7):1175–1194.

Hansen, W. G. 1959. How accessibility shapes land use. *Journal of the American Institute of Planners* 25 (2):73–76.

Hillier, B. 2007. *Space is the Machine: A Configurational Theory of Architecture.* London, UK: Space Syntax.

Hillier, B. and Hanson, J. 1989. *The Social Logic of Space.* Cambridge, UK: Cambridge University Press.

Ibeas, A., Lastra, J. M. D. P., Moura, J. L., Velasco, R. D. and Vega, A. 2002. Variaciones en la accesibilidad y en la renta por la puesta en servicio de la autovía de la Meseta. V Congreso de Ingeniería del Transporte, Santander.

Koenig, J. G. 1980. Indicators of urban accessibility: Theory and application. *Transportation* 9 (2):145–172. doi:10.1007/BF00167128.

McFadden, D. 1980. Econometric models for probabilistic choice among products. *Journal of Business* 53 (3):13–29.

Waddell, P., Ulfarsson, G. F., Franklin, J. P. and Lobb, J. 2007. Incorporating land use in metropolitan transportation planning. *Transportation Research Part A: Policy and Practice* 41 (5):382–410.

4 Microeconomic Theory of the Interaction between Transport and Land Use

Rubén Cordera, Ángel Ibeas and Borja Alonso

CONTENTS

This chapter will provide an introduction to the microeconomic models that have attempted to describe and explain the different patterns of population and activities in urban and regional systems. The models presented here are strongly linked to economic theory and they not only explain the location patterns of diverse activities but also the mechanism through which rural and urban land rents are established. This mechanism is based on a bidding process that is followed by households and companies in different locations according to their own budgetary constraints and location and consumer preferences.

Section 4.1 introduces the von Thünen agricultural location and land use model. This model is considered to be the first to introduce a microeconomic viewpoint based on the maximisation of utility by the agents involved in the land market. The von Thünen model has also been referenced by researchers later thanks to its rigor and analytical simplicity. Since the 1960s, the von Thünen model has been adapted to explain urban land use, the distribution of different social groups according to budgetary constraints and land rents resulting from the process. The basic Alonso (1964) urban location model has seen several later additions that have tried to explain new aspects such as the influence of time restrictions or the workings of the housing sector. Section 4.4 addresses the inter-urban models of Christaller and Lösch, which also have had a wide influence on the spatial modelling of market areas. Finally, Section 4.5 summarises the impact of these theories on the application of land use–transport interaction (LUTI) models to urban planning. This is an impact that has always been limited given the dif-ficulties these models present to professionals when they try and apply abstract ideas that are based on simplified hypotheses, which are difficult to adapt to real-planning situations.

4.1 THE VON THÜNEN AGRICULTURAL LAND USE MODEL

4.1.1 BASIC ASSUMPTIONS

The microeconomic land use theory formulated by von Thünen (1826) was original in the sense that it was the first attempt to incorporate the effect of transport costs into an explanation of rents and agricultural land use patterns. In its simplest form, the model assumes the following basic hypotheses (Camagni 2005):

- A closed and an idealised agricultural region with identical unit transport costs, soil fertility and production factors in all directions.
- One central market where all the producers go to sell their produce.
- One unique production function for each type of agricultural product with constant returns to scale.
- Prices for each product provided exogenously.
- Unlimited demand for products.

According to the model, the farmers compete for the different plots of land and the one who is prepared to pay the highest rent takes over the plot. The rent emerges from the production process as a residual after removing the transport costs, the typi-cal profit for that particular product and the production costs from the farmer's total income. As the profits and production costs are already given, the rent will depend on the transport costs and therefore the nearest farmers to the central market will be able to pay the higher rents.

Mathematically, for any product m, the maximum unit rent will be equal to

$$S_j^m = q^m(p^m - c^m - k^m d_{ij})$$
(4.1)

TABLE 4.1
Bid Rent According to Transport Costs

Distance to the Central Market	Price	Quantity Produced	Overall Profit	Unit Production Cost	Unit Transport Cost	Bid Rent
Area close to the market	10	4	40	4	1	20
Area away from the market	10	4	40	4	3	12

where:
S_j^m is the rent or surplus for product m at location j
q^m is the amount of product m produced per unit of land
p^m is the price per unit of product m
c^m is the cost of production per unit of product m
k^m is the unit transport cost
d_{ij} is the distance between location j and the central market i

Equation 4.1 is a linear function in which the maximum is given when the transport costs are zero; in other words, when the location is the central market. The minimum comes when $k^m d_{ij}$ is equal to $p^m - c^m$, representing a location where the transport costs are greater than the profits obtained, meaning that the agricultural production is not profitable.

As an example, let us assume an agricultural area with two large zones: (1) one near to the central market and (2) another much further away. Both zones produce the same product, wheat, for the same market price. The farmers who are further away could pay lower rents because of their greater transport costs (Table 4.1). This process also shows that the land rents received by the land owners mean that none of the farmers will have excessive profits greater than the amount that will allow them to continue producing.

4.1.2 CALCULATING THE DISTRIBUTION OF LAND USE AND LAND RENTS

Given that the model contemplates more than one product m, von Thünen assumes that the land owners will rent the land to the farmers who make them the highest offers. This supposes that the farmers compete for the locations and all have the ability to bid the maximum allowed by the surplus obtained from product m at location j. As can be seen in Figure 4.1, the competition between different farmers (in example three) and the assignment of land to those farmers who can offer a higher rent, given their larger surplus, results in a concentric land use pattern. As the order is given by $q^m (p^m - c^m)$, those farmers with a greater differential between prices and costs and with more intensive production methods per unit of land, will be able to offer higher land rents and locate closer to the central market. From this point, the surplus

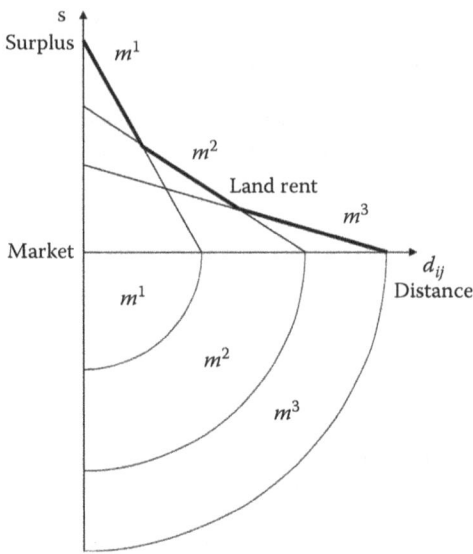

FIGURE 4.1 Rent and spatial distribution of three agricultural products according to the von Thünen model.

will fall as a function of k^m; in other words, the unit transport costs for each type of product. The land rents will be determined by the highest rents being offered for each type of product highlighted in Figure 4.1. A numerical example can be found in Table 4.2 for the three agricultural products: (1) intensive cultivation, (2) extensive cultivation and (3) forest.

The model can be made more realistic by incorporating a more complex function for surplus, for example, by reducing transport costs with distance or assuming elasticity in the demand compared to the price of goods (Barra 1989). Furthermore, the land use patterns can also be made more complex by introducing various central markets in the same study area.

TABLE 4.2
Bid Rent for Three Agricultural Products

Product	Price (€)	Quantity Produced	Total Profit (€)	Unit Production Cost (€)	Unit Transport Cost (€)	Bid Rent (€)
m^1– Intensive cultivation	10	10	100	5	1	40
m^2– Extensive cultivation	5	16	80	2	2	16
m^3– Forest	3	18	54	0.5	2.4	1.8

4.2   ALONSO'S MONOCENTRIC CITY MODEL

The basic ideas contained in the model formulated by von Thünen can be adapted to the context of urban location. This line of research was followed by Wingo (1961) and Alonso (1964) to relate transport costs with the location of activities and urban land use. Both models have a lot of similarities, although Alonso's model has had the greater influence and has inspired later developments.

4.2.1   ALONSO MODEL OF BID–RENT THEORY

The simplified hypotheses of the Alonso model are similar in idea to those of von Thünen for the case of agricultural land use:

- A closed urban area with homogenous space in all directions and identical unit transport costs without the presence of any kind of congestion.
- A single commercial and business centre (CBD) where all the jobs in the urban area are located and where all the companies wish to locate, given that it is the centre of exportation and information exchange.

Taking on board both suppositions, the only variable that differentiates each location is the cost of travelling to the CBD. The Alonso model considers households as land users and differentiates them by rent, whereas companies are differentiated by the economic sector. Each agent has a budgetary constraint represented by the sum of the transport costs, land rent and an expense on all other goods, shown by

$$y = r_j l_j + c_{ij} + p_z z \tag{4.2}$$

where:
   y is the income of the household or company
   r_j is the rent per unit of land
   l_j is the amount of land
   c_{ij} are the transport costs between i and j
   p_z is the unit price of the composite good
   z is the number of units of the composite good consumed

The three elements (Equation 4.2) result in a three-dimensional surface with all the possible combinations of expenses providing a determined level of utility. The model also assumes the maximisation of this utility by the urban agents. The rent offered by each agent per unit of land can be found by transforming Equation 4.2, which becomes

$$r_j = \frac{y - c_{ij} - p_z z}{l_j} \tag{4.3}$$

Graphically, if only the bid rent and the distance to the CBD are considered (Figure 4.2), each household or company can have multiple curves representing the bid rent on which each point generates the same utility. The lowest curves are

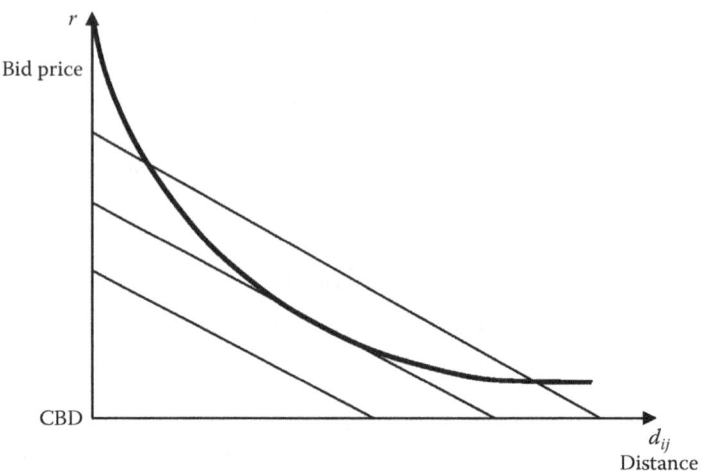

FIGURE 4.2   Bid–rent curves in the Alonso model.

the most desirable as they offer the same level of utility at a lower cost. However, the effective rent only occurs at the point where the lowest bid curve crosses the effective rent curve (highlighted in bold in Figure 4.2), according to the market. At this point, the slopes of both curves are equal, in other words, $dr/dc = dr'/dc$ where c is the expense on transport and r' is the effective market rent. Another important conclusion from the model is that r_j should fall as the transport costs rise in order to maintain a given level of utility. Therefore, the model predicts the existence of a negative gradient in the unit land prices moving away from the CBD, a fact with evidence in multiple study areas, especially when they are of a monocentric nature as is assumed by the model (Glaeser 2008).

Table 4.3 presents an example of location choice made by a household with an income of €2000. The household is able to locate in the furthest zone from the CBD with the same level of utility, and the greater transport and consumer costs

TABLE 4.3

Rent Offered by a Household at Different Distances from the CBD for the Same Level of Utility

Distance from the CBD	Household Budget (€)	Unit Price of Land (€)	Amount of Land Consumed (m²)	Unit Price of Transport (€)	Distance (m)	Spending on Other Goods (€)
Area closest to the CBD	2000	5	50	0.5	500	1500
Area furthest from the CBD	2000	3	100	0.5	1000	1200

are compensated by having more land. Another possibility is to choose somewhere closer to the CBD with lower transport costs but with the implication of a higher rent. In this case, in order to maintain the level of utility, the household can obtain less land, and use the remaining income to consume other goods. Therefore, the household will find location equilibrium in both zones as they will both provide an identical level of utility.

4.2.2   CALCULATING THE DISTRIBUTION OF LAND USE AND RENTS FOR URBAN LAND

In the case where there are various kinds of users, the slopes of the curves representing the bid rent will vary for each of them due to their different preferences in relation to the transport costs and their substitution for z and l. For example, a user or a company with a high preference for locating in the centre will show a bid–rent curve with a steeper slope. If all the location choices of all the companies and households are seen at the same time, it becomes possible to do it without an effective rent function as provided in the previous case and substitute it by the envelope of the bid–rent curves of all the agents seeking to locate. Alonso proposed a solution algorithm which, starting at the CBD, located the different urban agents from higher to lower in accordance with the bid–rent slope functions. Once the last user was located, the land rent just outside the limits of the urban area corresponded to the price for agricultural land (Figure 4.3).

In a similar way to the von Thünen model, the land uses will show a concentric spatial pattern around the CBD. Each location will be occupied by the user who offered the highest rent and the different types of activities and households will have indifference in the localization within each concentric area. Furthermore, the location will be stable because the spatial pattern is optimal considering the preferences and financial constraints of the agents.

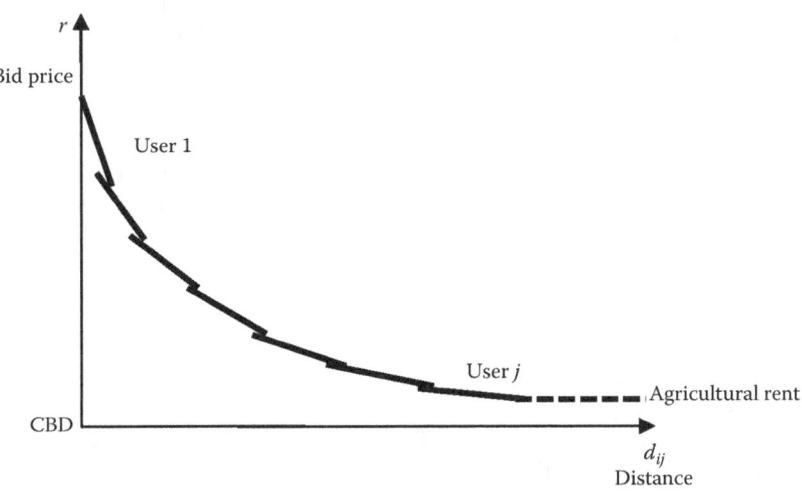

FIGURE 4.3   Envelope of the bid–rent curves for j users.

4.3   LATER ADDITIONS TO THE ALONSO MODEL

The Alonso model has been expanded in different directions to form the theoretical nucleus of the modern urban economics. Two of the more widely recognised contributions correspond to the works of Muth (1969) and Beckmann (1974). Both cases keep the basic assumptions of the Alonso model, constructed around the supposition of a balance between the costs of transport to the CBD and the unit price of land. The spatial equilibrium between the different uses occurs because the more distant activities and households having the highest transport costs are compensated by lower land rents. Differently from Alonso, however, the models of Muth and Beckman add realism by considering, in the case of Muth, that land itself is not the final consumer product but is an intermediary for the consumption of the final product, which is the home. Beckman added commuting time to the CBD to his model along with a time constraint, which the urban agents had to fulfil.

4.3.1   THE MUTH MODEL

The Muth model extends the Alonso model by incorporating the consumption of an aggregated product known as housing services. This leads to the budgetary restrictions of each agent being represented by

$$y = r'_j q_j + c_{ij} + p_z z \tag{4.4}$$

where:
   r'_j is the unit housing service price in location j
   q_j is the amount of housing service consumed

In addition, a function of the housing services production generated by the industry is also specified. This production function depends on the land L and the capital K invested. So, the industry in each location j tries to maximise

$$P = r'_j F(L,K) - r_j L - K \tag{4.5}$$

where P is the profit and r_j the land rent in location j. Equations 4.4 and 4.5 can be combined to obtain a location choice model in which each household chooses the desired inputs of land and capital as in Fujita (1989):

$$y = r'_j l_j + c_{ij} + p_z z + p_k k \tag{4.6}$$

This model is identical to the basic model Equation 4.2 except for the inclusion of k, the amount of capital chosen by the household and p_k the price of the capital. Therefore, the capital–land ratio in the production of housing services falls with the distance from the CBD meaning that the density of the built space drops as distance to the CBD increases, an empirical fact that is common to many urban areas.

4.3.2 THE BECKMAN COMMUTING TIME MODEL

The model developed by Beckman explicitly introduces the commuting time to the CBD by considering it as important as the monetary cost of the journey. The households will maximise the utility of their location choice subject to a budgetary and a time constraint:

$$y_{nw} + Wt_w = r_j l_j + ad_{ij} + p_z z$$

$$t = t_w + t_l + t_c d_{ij}$$

(4.7)

where:

y_{nw} is the non-wage income

Wt_w is the wage income where W is the salary rate

$r_j l_j$ is the land rent

ad_{ij} is the cost of transport between i and j where a is the unit distance cost

t is the total available time

t_w is the time spent on working

t_l is the leisure time

$t_c d_{ij}$ is the commuting time between i and j where t_c is the commuting time per
  unit of distance

The potential wage income in a location j may also be specified as

$$I_{jw} = W(t - t_c d_{ij})$$

(4.8)

And the net potential income as

$$I_j = y_{nw} + I_{jw} - ad_{ij}$$

(4.9)

Furthermore, t_w can be expressed as $t_w = t - t_l - t_c d$, which means the location choice model can now be presented as a process of maximising the utility subject to only one budgetary constraint:

$$I_j = Wt_l + r_j l_j + p_z z$$

(4.10)

where Wt_l can be interpreted as the purchase of leisure time after having sold all the available time minus the commuting time at salary rate W. Therefore, salary rate W also plays the role of the unit price of leisure time (Fujita 1989). Finally, the rent offered by the agents for a location j can be expressed as

$$r_j = \frac{I_j - Wt_l - p_z z}{l_j}$$

(4.11)

This model shows that the households with the lowest incomes prefer to have lower commuting times, when this time implies an important loss of potential wage income and potential net income as this kind of households normally has zero nonwage incomes. This is a typical spatial location pattern that can be found in the cities of the United States, where the lower income households occupy the land close to the CBD. For middle income households, the importance of commuting time will be lower because they tend to occupy locations further away from the CBD. LeRoy and Sonstelie (1983) contributed further evidence about the different modes of transport used by households with different incomes. While the lower income households tended to use public transport more, the middle and higher income households tended to use private transport and therefore showed lower commuting times per unit of distance. The conclusions of the model are further reinforced in the case of considering two available modes of transport. However, when the salary rates are very high, the model predicts that households once again prefer to locate close to the CBD to avoid long commuting times and loss of income.

4.4   THE CHRISTALLER AND LÖSCH MODEL OF MARKET AREAS

The urban economic models presented until now try to simulate the basic mechanisms explaining the spatial pattern of land uses as well as the agricultural and urban rents that emerge from the location process in a local territorial system. Other models, however, have focussed on trying to simulate the distribution pattern of the urban and the rural nuclei themselves and the market areas that these locations generate around them. Empirical observation shows how towns and cities and rural settlements form a hierarchy, where the smallest centres adopt limited economic functions, whereas the larger centres take on more functions, some of which are highly specialised. The theories of Christaller and Lösch try to answer questions such as

- How is the hierarchy between the settlements produced?
- What is the frequency of each hierarchical level?
- What is the size of the market area served by each hierarchical level?
- What is the spatial distribution pattern of the resulting settlements?

4.4.1   THE SPATIAL DEMAND CURVE AND SPATIAL MARKET EQUILIBRIUM

Before explaining the main characteristics of the Christaller and Lösch models, it would be beneficial to discuss the concepts of the spatial demand curve and spatial market equilibrium. The spatial demand curve is a conventional demand curve, which explicitly considers the location of production and consumers in a territory. The producers in a determined settlement can sell their produce in the surrounding area where the transport costs will progressively increase the price of the goods, so the demand will progressively drop until it becomes null. Therefore, this spatial demand curve will aggregate the demand of all the local consumers. If this demand curve rotates 360°, the so-called Lösch demand cone is generated, which depends on the consumer preferences as well as the transport costs and the price of production at origin, as can be seen in Figure 4.4. The concept of the spatial demand curve is also related to the concepts of a product's range and threshold. The range of a product is the maximum distance over which it can be sold, shown by the radius of the demand cone. This concept is related to a product's threshold, or the minimum distance required so that the production of a product is profitable when supplying a

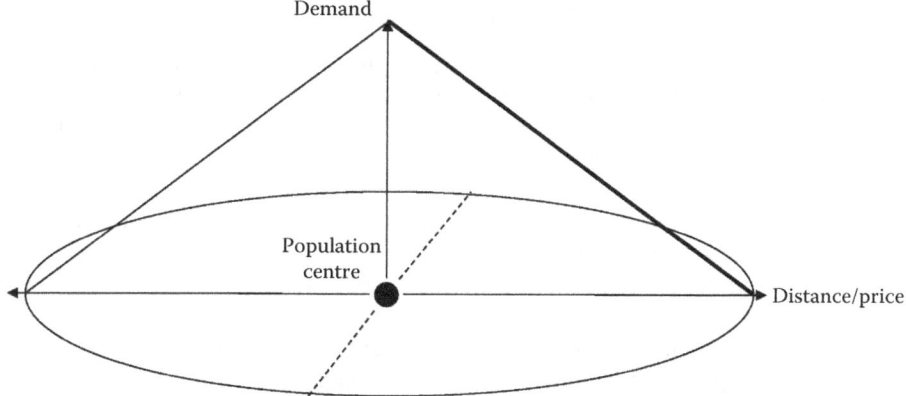

FIGURE 4.4   Demand cone around a settlement.

sufficient and effective amount of demand. Therefore, a product's threshold must be lower than its range in order for it to be produced at a given settlement.

When the market areas of more than one settlement over cross each other, then both will contract sharing the demand as a function of the distance to the closest settlement. As more settlements start joining the idealised region, the market areas of the settlements will get smaller and smaller until the limit where the price obtained is equal to the cost of production plus the normal profit from creating that product. This means that the settlements will have reached spatial market equilibrium. The settlements will be as close as possible and the resulting spatial pattern will form a continuum of hexagonal market areas around the settlements, as shown in Figure 4.5.

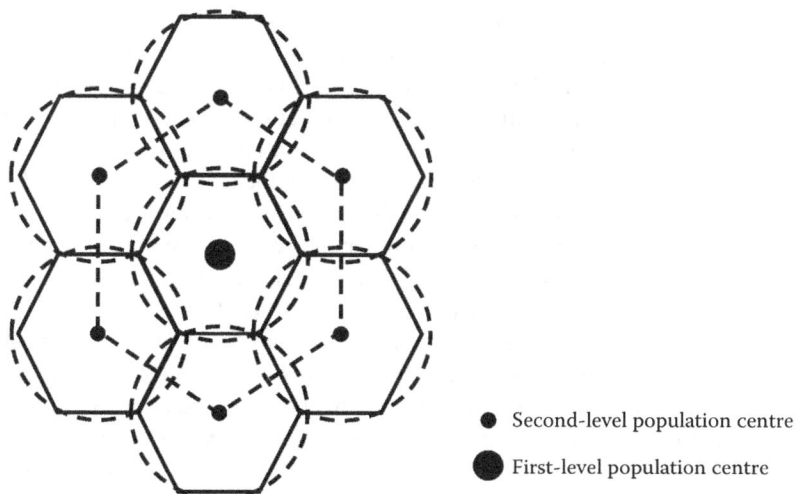

● Second-level population centre

⬤ First-level population centre

FIGURE 4.5   Superimposed market areas around central places with the resulting hexagonal spatial pattern.

4.4.2   THE CHRISTALLER CENTRAL PLACE MODEL

The model proposed by Christaller (1933) represents the first geographical approach to explain the problem of central places, that is, the population nuclei understood to be the providers of goods and services to the surrounding areas. The simplified hypotheses of the model are as follows:

- A closed regional area with a homogenous distribution of a population made up of self-sufficient farmers
- An isotropic space with identical transport costs in all directions

In the Christaller model, certain settlements are allowed to concentrate on the production of more specialised goods with a larger demand threshold. This will generate larger market areas around these settlements and also lead to the hierarchisation of the settlements. The first-level settlements will be those that concentrate the production of a greater number of goods and services. The market area surrounding the first-level settlement will contain its own lower level market area within the hexagonal structure plus 6/3 market areas of the nearby settlements, representing a total of 3 lower level market areas, as can be seen in Figure 4.5. This number of market areas contained in an area's upper hierarchical level is given by the value K in the Christaller theory. Each step in the hierarchy corresponding to a K value presents $N_1 = K^1$ number of places where l is the step in the hierarchy. So, for example, for the fourth hierarchical step in a $K = 3$ hierarchy, the central place would have $3^4 = 81$ lower level market areas or dependent settlements within its sphere of influence or market area.

Christaller called this mechanism for structuring the system of settlements with $K = 3$ market principle as it is based on the minimisation of the distance between higher level places and their dependent settlements. Two additional hierarchical principles were also proposed. First, a transport principle in which each upper market area includes the surface area of $K = 4$ lower market areas as the dependent settlements are located on the intersections of the transport network between the upper level settlements in such a way as to minimise the cost of the network. Second, the principle of administrative organisation, where each upper level market area includes $K = 7$ lower level market areas as they are completely contained in the former. Figure 4.6 shows a representation of the spatial organisation of the three cases. As can be seen, each secondary settlement in the $K = 3$ organisation depends on the three upper level settlements, whereas it is only two in the $K = 4$ system. Finally, for $K = 7$, all the secondary level market areas are completely contained in the first-level areas.

Christaller applied his model to the empirical analysis of an area in Southern Germany and obtained quite a good fit as the reality of hierarchical levels and the number of observed central places compared well to the theory's prediction. Several empirical tests were done later, such as the study of the county of Snohomish by Berry and Garrison (1958) who found a significant fit with the observed data. More recently, the theory has continued to be applied to cases such as the industrial activities (Mori et al. 2008) who also found their distribution had a correct fit with the stepped hierarchy

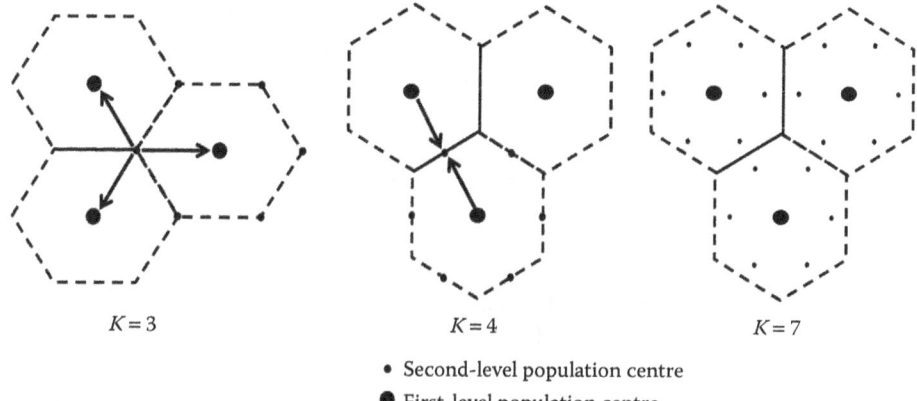

$K = 3$ $K = 4$ $K = 7$

• Second-level population centre

⬤ First-level population centre

FIGURE 4.6   Organisation of the market areas $K = 3$, $K = 4$ and $K = 7$ according to Christaller central place theory.

derived from the central place theory. However, the theory has also been exposed to multiple criticisms precisely due to the deduction of a stepped-scale population pattern. Generally, the empirical studies that have quantified the relationship between the number of functions a settlement performs and the population it serves have found continuous distributions, which have required additional classification techniques to establish the hierarchical steps deduced from the Christaller model (Beavon 1977).

4.4.3   The Lösch Model of Multiple Market Areas

August Lösch (1954) expanded the Christaller model and made it more flexible but kept the hexagonal structure of the market areas due to spatial-economic equilibrium. However, Lösch broke with the imposition that the K value should be fixed for all the hierarchical levels in the system of central places as well as with the hypothesis that each centre should adopt all the economical functions of the lower level centres. Lösch considers settlement hierarchies and uses K values just similar to those proposed by Christaller (3, 4, and 7); however, he also adds $K = 9$, $K = 12$, $K = 13$ and other higher levels up to $K = 511$ representing the market area of the highest level settlement. In all cases, the greatest centre serves all the goods that are present in the system, whereas the rest of the hexagonal areas for all the K values rotates around the axis of this main centre until the greatest possible superimposition of centres is reached. By using this method, Lösch was able to generate a system of settlements with a continuous hierarchy, instead of the discrete steps coming from the Christaller model. Furthermore, the pattern of central places generated by the Lösch model is not spatially homogenous but radially unequal, with some areas having greater settlement densities with a greater presence of larger settlements and areas with fewer and smaller settlements. These areas with differential patterns of settlement density were denominated by Lösch as rich zones and poor zones.

The continuous hierarchy of settlements generated by the Lösch model generally has a better fit to the rank-size rule based on a log-normal distribution, which usually replicates empirical distributions with a higher fit (R^2 greater than 0.9) (Beavon 1977). This rank-size rule takes the following form:

$$\log P_r = \log P^* - \beta \log r \qquad (4.12)$$

where:
   P_r is the size of a settlement for a rank r
   P^* is the size of the biggest settlement
   β is a parameter to be estimated

As Equation 4.12 is linear in the parameter β, it can be directly estimated using ordinary least squares.

4.5 SOME CONCLUSIONS ABOUT MICROECONOMIC MODELS AND THEIR APPLICATION TO LAND USE–TRANSPORT INTERACTION MODELS

The models presented in this chapter are coherent with microeconomic theory based on the maximisation of utility by agents and the solution of their interaction through finding the market equilibrium. The resulting spatial pattern and land rents are derived deductively from these principles in coherence with economic theory. However, the application of these models in planning has been relatively limited due to various difficulties.

4.5.1 SOME CONSIDERATIONS ABOUT THE PROBLEMS FOUND USING URBAN ECONOMIC MODELS IN THE APPLIED FIELD

In spite of the theoretical coherence and the parsimony provided by the urban location models and hierarchical central place system, they have not been directly inspirational in the creation of many practical, operational and applied models based on their formulation. Due to characteristics such as the treatment of space as a continuous variable, these models have turned out to be too general for addressing discrete and particular characteristics in urban areas, which quite often decisively condition urban location. Furthermore, the original Alonso model starts from the hypothesis of a single unique urban centre, which has been shown to be increasingly restrictive, given that many urban areas have evolved by generating important secondary centres (Garreau 2011). However, this hypothesis has been relaxed by authors such as Henderson and Mitra (1996) who introduced multiple centres of employment into the Alonso–Muth model. The relationship shown by these models between the distance to the urban centre and the drop in real-estate prices has become null because households and activities can take advantage of lower transport costs to secondary centres. This fact has been empirically corroborated in the case of the United States (Glaeser et al. 2001). Wheaton (1977) also performed an empirical study with crossed data for the United States and showed that the willingness to pay for land is not significantly different between the groups of different income levels, where the Alonso–Muth model would not have the capacity to explain the spatial location choices of households with different incomes. These choices would better be explained by individual factors such as environmental externalities or the different tax burdens between the city centre and the suburbs. In the case of Europe, the urban location models have also encountered problems when it comes to their practical application because historical town centres have continued to be occupied by high income households. The location patterns in Europe have therefore been notably different from the more frequently found patterns in North America, which were taken as references when formulating the model.

In spite of the earlier, this type of model derived from microeconomic theory has contributed an analytical tool, which allows us to understand some of the basic mechanisms, which explain the urban phenomena. In this sense, they have been and continue to be useful by providing a general framework for the functioning of urban location and therefore also providing hypotheses that can be checked and tested when compared to the reality to guide empirical studies. As stated by Glaeser (2007), the Alonso and Muth models made it possible to make generalised predictions in order to compare a variety of urban areas rather than concentrate on the particularities of different places. They were also coherent with the economic theory as their starting point was the idea of a spatial equilibrium similar to the idea of market equilibrium in the wider economy or the absence of arbitrage in the financial sector.

4.5.2 Applications of Microeconomic Models in LUTI Models

One of the more notable exceptions to the scarcity of practical applications derived from models of urban economic theory is the Herbert and Stevens (1960) optimisation model. This model adapts Alonso's urban location theory by formulating it as an optimisation problem for maximising rents for urban land subject to restrictions such as the available space in each area. More details about this particular model can be found in Chapter 7. The Herbert–Stevens model was applied to the Penn–Jersey area as part of the Philadelphia metropolitan transport plan. However, the model only considered the residential real-estate market and did not address the supply of a built space, nor did it consider any kind of environmental externality; it also had a deterministic nature, which removed a high degree of realism. This has limited its applications in spite of the proposal to make later improvements such as the addition of a stochastic component (Grigg 1984) or of a transport model (Los 1979).

Martinez (1992) also adapted Alonso's bid–rent model using the hypothesis of land assignment according to the highest bid made by the urban agents. The willingness to pay for each place is calculated by the Martinez model as a function of both the attributes of the plots and the agents who are bidding, also adding an identically distributed and independent Gumbel-type error term, which leads to an expression of location choice using a Multinomial Logit type of discrete choice model (see Chapter 6). This model has been applied to the urban area of Santiago (Chile) where it was used to simulate different urban planning scenarios.

REFERENCES

Alonso, W. 1964. *Location and Land Use: Toward a General Theory of Land Rent, Publications of the Joint Center for Urban Studies of the Massachusetts Institute of Technology and Harvard University*. Cambridge, UK: Harvard University Press.

Barra, T. 1989. *Integrated Land Use and Transport Modelling: Decision Chains and Hierarchies, Cambridge Urban and Architectural Studies*. Cambridge, UK: Cambridge University Press.

Beavon, K. S. O. 1977. *Central Place Theory; A Reinterpretation*. London, UK: Longman Publishing Group.

Beckmann, M. J. 1974. Spatial equilibrium in the housing market. *Journal of Urban Economics* 1 (1):99–107.

Berry, B. J. L. and Garrison, W. L. 1958. The functional bases of the central place hierarchy. *Economic Geography* 34 (2):145–154.

Camagni, R. (Ed.). 2005. In *Economía Urbana*, Bosch, A. (Ed.). Barcelona, Spain: Antoni Bosch.

Christaller, W. 1933. *Die zentralen Orte in Süddeutschland*. Jena, Germany: Gustaf Fisher. Translated by Carlisle, W. B. (1966), as *Central Places in Southern Germany*. Englewood Cliffs, NJ: Prentice Hall.

Fujita, M. 1989. *Urban Economic Theory: Land Use and City Size*. Cambridge, UK: Cambridge University Press.

Garreau, J. 2011. *Edge City: Life on the New Frontier*. New York: Anchor.

Glaeser, E. L. 2007. The economics approach to cities. In *NBER Working Paper Series Working Paper 13696*. Cambridge, MA: National Bureau of Economic Research.

Glaeser, E. L. 2008. *Cities, Agglomeration, and Spatial Equilibrium, Lindahl Lectures*. Oxford, UK: Oxford University Press.

Glaeser, E. L., Kahn, M. E., Arnott, R. and Mayer, C. 2001. Decentralized employment and the transformation of the American city. *Brookings-Wharton Papers on Urban Affairs*, pp. 1–63.

Grigg, T. J. 1984. Probabilistic versions of the short-run Herbert—Stevens model. *Environment and Planning A* 16 (6):715–732. doi:10.1068/a160715.

Henderson, V. and Mitra, A. 1996. The new urban landscape: Developers and edge cities. *Regional Science and Urban Economics* 26 (6):613–643.

Herbert, J. D. and Stevens, B. H. 1960. A model for the distribution of residential activity in urban areas. *Journal of Regional Science* 2 (2):21–36. doi:10.1111/j.1467-9787.1960.tb00838.x.

LeRoy, S. F. and Sonstelie, J. 1983. Paradise lost and regained: Transportation innovation, income, and residential location. *Journal of Urban Economics* 13 (1):67–89. doi:10.1016/0094-1190(83)90046-3.

Los, M. 1979. Combined residential-location and transportation models. *Environment and Planning A* 11 (11):1241–1265. doi:10.1068/a111241.

Lösch, A. 1954. *The Economics of Location*. New Haven, CT: Yale University Press.

Martinez, F. J. 1992. The bid-choice land-use model: an integrated economic framework. *Environment & Planning A* 24 (6):871–885.

Mori, T., Nishikimi, K. and Smith, T. E. 2008. The number-average size rule: a new empirical relationship between industrial location and city size. *Journal of Regional Science* 48 (1):165–211.

Muth, R. F. 1969. *Cities and Housing; The Spatial Pattern of Urban Residential Land Use*. Graduate School of Business, University of Chicago. Third Series: Studies in Business and Society. Chicago, IL: University of Chicago Press.

von Thünen, J. H. 1826. *Der isolierte staat in beziehung auf landwirtschaft und nationaloekonomie.* Jena, Germany: Gustaf Fisher. Translated by Wartenburg, C. M. (1966), as *The Isolated State.* Oxford, UK: Oxford University Press.

Wheaton, W. C. 1977. Income and urban residence: An analysis of consumer demand for location. *The American Economic Review* 67 (4):620–631.

Wingo, L. 1961. *Transportation and Urban Land.* Baltimore, MD: Johns Hopkins Press.

5 Spatial Interaction Models

Gonzalo Antolín, Rubén Cordera and Borja Alonso

CONTENTS

5.1 THE ORIGINS OF SPATIAL INTERACTION MODELS

Spatial interaction models simulate the journeys made between origins and destinations, for example, between the land uses found in the different zones in an urban or regional system. These interactions can be displacements on any scale such as international, national, regional or local and they could be made for multiple reasons such as work, study, shopping or leisure.

The first research in the field of spatial interaction modelling dates from the 1930s and later; examples are the work of Reilly (1931), Hoyt (1939), Stewart (1948), Zipf (1949), Converse (1949), Clark (1951) and Isard (1956). However, most authors coincide in that these models were positioned within the wider framework of land use–transport interaction (LUTI) modelling in the work of Hansen (1959), who used concepts derived from spatial interaction theory to simulate the location of residents within an urban context. Important progress was later made with interaction models in research related to transport, mainly because their application to real cases is relatively simple and good fits were generated with observed journeys.

In principle, spatial interaction models were developed using an aggregated methodology that is based on an analogy with the Newtonian law of gravity. One of the first steps in this direction was the work of Hansen (1959) mentioned earlier, who used the analogy with gravity to simulate the location of residents whilst considering the accessibility to employment and the available land in each area. Later, Huff (1963) made an important contribution by interpreting the basic gravity model in terms of economics and probabilities. Lowry (1964) reached another important milestone by relating this type of model with the theory of economic base when providing an explanation of economic-spatial dynamism within urban areas. The Lowry model was improved later by Garin (1966) and Rogers (1968) using matrix estimation techniques. Alan Wilson (1970, 1974) provided an alternative theoretical foundation for spatial interaction models, which went further than the gravity analogy by using the concept of maximum entropy. All these developments led to the creation of the first operational LUTI spatial interaction models reflected in works such as those of Echenique (1971) and Batty (1976).

5.2   BASIC CONCEPTS OF SPATIAL INTERACTION MODELS

Spatial interaction models have an aggregated nature because both space and the traveller characteristics can be grouped into large categories. So, instead of studying determined points in space, larger zones are defined containing many activities and occupations in which the number of journeys being produced and attracted is analysed. Furthermore, people are aggregated into groups where the members share similar socioeconomic and behavioural characteristics. Figure 5.1 presents an example of an urban area that has been divided into four large areas. The arrows on the diagram represent the flows of people moving between the zones. These are the flows that interaction models are trying to simulate by fitting the predictions as well as possible to the observed data.

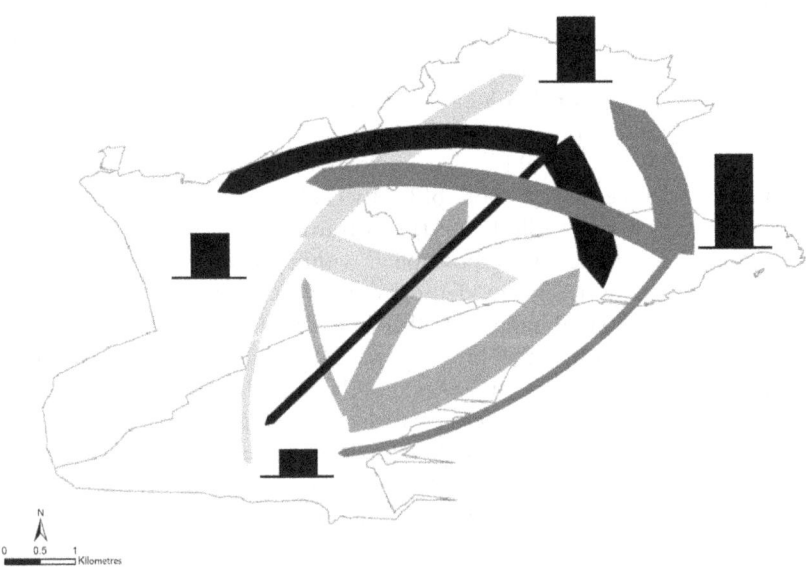

FIGURE 5.1   Journeys between zones within an urban area.

5.2.1 THE GRAVITY MODEL

The analogy with gravity in the framework of spatial interaction models assumes that the interaction between the zones is proportional to the number of people or activities present in each area and is inversely proportional to the friction imposed by the travelling that has to be made between them.

The first use of a gravity model within a framework of urban mobility was made by Casey (1955), who applied this approach to simulate shopping trips made between the regional towns. The initial formulation of the model was as follows (Ortúzar and Willumsen 2011):

$$T_{ij} = \frac{\alpha P_i P_j}{d_{ij}^2} \tag{5.1}$$

where:

T_{ij} are the journeys between origin i and destination j

P_i and P_j are the populations of the origin and destination cities for the trips

d_{ij} is the distance between both zones

α is a proportionality factor, which allows the units of population to be transformed into trips

This formulation was later thought to be an analogy with the law of gravity too simple, so the overall generated and attracted trips (O_i and D_j) were used instead of the populations of the zones. It was then considered that the effect of distance could be modelled more efficiently using a decreasing function according to the distance or cost. After making these corrections, the resulting model was as follows:

$$T_{ij} = \alpha O_i D_j f(c_{ij}) \tag{5.2}$$

where $f(c_{ij})$ is the generalized function of journey cost, which has one or several parameters to be calibrated. This function, generally called the *trip resistance function* or *friction function*, represents the cost or resistance of making a journey that increases with the distance, time or the cost of the journey. The cost functions can take different forms (exponential, potential and combined) and will be developed in greater detail in Chapter 11.

If the trips are represented in a matrix, then in the context of the gravity models different types of models can be defined according to the constraints they consider. Two of the possible constraints to consider are

$$\sum_j T_{ij} = O_i \tag{5.3}$$

$$\sum_i T_{ij} = D_j \tag{5.4}$$

These constraints show us that the sum of the trips on each of the matrix rows should be equal to the total number of trips generated by the zone that the row refers to; analogously, the sum of the trips for each matrix column should correspond to the number of trips attracted by the zone the column refers to. For both constraints to be fulfilled, the proportionality factor α needs to be replaced by the two balancing factors A_i and B_j. When these substitutions are put into place, the gravity model adopts the following expression:

$$T_{ij} = A_i O_i B_j D_j f(c_{ij}) \tag{5.5}$$

Expression 5.5 represents the classic version of the doubly constrained gravity model. The singly constrained versions of the model, both in origins and in destinations, can be derived by equalling one of the balancing factors B_j or A_i, respectively. In the case of a destination constrained model this would be $A_i = 1$, meaning

$$B_j = \frac{1}{\sum_i O_i f(c_{ij})} \tag{5.6}$$

Whereas in the case of a doubly constrained model, the balancing factors are

$$A_i = \frac{1}{\sum_j B_j D_j f(c_{ij})} \tag{5.7}$$

$$B_j = \frac{1}{\sum_i A_i O_i f(c_{ij})} \tag{5.8}$$

These balancing factors are interdependent, which is why in the case of the doubly constrained model the values of both groups are required. This is done by performing an iterative process initiated by making all $B_j = 1$, then by considering $f(c_{ij})$ and calculating A_i. Following that, the B_j are once again calculated with these values and the steps are repeated until the process converges at a solution or it gets close enough to a defined stop criterion, which fits the observed data sufficiently well to maintain the trustworthiness of the model.

5.2.2 THE CONCEPT OF MAXIMUM ENTROPY

The concept of maximum entropy has been applied to a generation of numerous interaction models such as residential location models, shopping models or the doubly constrained gravity model. Wilson (1970) introduced the maximum entropy method and gave the spatial interaction models a whole new theoretical framework in which to develop.

The context of maximum entropy considers a system formed of a high number of differentiated elements. A description of each one of the micro-states is required in order to provide an overall description. These are understood to be each individual person and the characteristics of their journey: origin, destination, mode of transport, journey time and so on. However, given that in many cases the amount of available information is insufficient, more aggregated information can be used; this is known as a meso-state. Each meso-state can be defined as a grouping of diverse micro-states with similar features such as the origin and destination.

A further body of information is available from the meso-states on an upper level in which the information is more aggregated; these are known as the macro-states. The macro-states are systems that contain information relative to the overall number of generated or attracted trips in a zone or the number of journeys on a link.

The maximum entropy method can be deduced from the description of the system made up of macro-states, meso-states and micro-states. Where no information to the contrary is available, it must be accepted that all the micro-states aggregated into meso-states or into macro-states having the same probability of existing. Therefore, according to Wilson (1970), the number of micro-states $W\{T_{ij}\}$ associated with the meso-state T_{ij} is given by the function:

$$W\{T_{ij}\} = \frac{T!}{\prod_{ij} T_{ij}!} \tag{5.9}$$

As all the micro-states are equally probable, the most likely matrix is the one that maximizes W, making the most probable meso-state the one with the highest number of the associated micro-states.

Applying logarithms to the formula W leads to

$$\log W = \log \frac{T!}{\prod_{ij} T_{ij}!} = \log T! - \sum_{ij} \log T_{ij}! \tag{5.10}$$

Applying the Stirling approximation ($\log X! = X \log X - X!$), we obtain (Ortúzar and Willumsen 2011):

$$\log W = \log T! - \sum_{ij} (T_{ij} \log T_{ij} - T_{ij}) \tag{5.11}$$

where the term $\log T!$ is a constant. However, it can be omitted in the final maximisation because it is the rest of the equation that defines the entropy function:

$$\log W' = -\sum_{ij} (T_{ij} \log T_{ij} - T_{ij}) \tag{5.12}$$

The maximisation of $\log W'$, subjected to the constraints, corresponding to the available knowledge about the macro-states, allows models to be generated to estimate the most probable meso-state, which in this case, corresponds to the most probable estimation of the T matrix.

If a simplified state is considered, it means that the micro-states of a macro-state can be calculated, as shown in Figure 5.2. This macro-state is formed of only two zones and four travellers in which the individuals a_1 and a_2 only travel from A and the individuals b_1 and b_2 only travel from B. The most probable meso-state is the first one in which the most micro-states are present with an individual making a journey in each origin–destination pair.

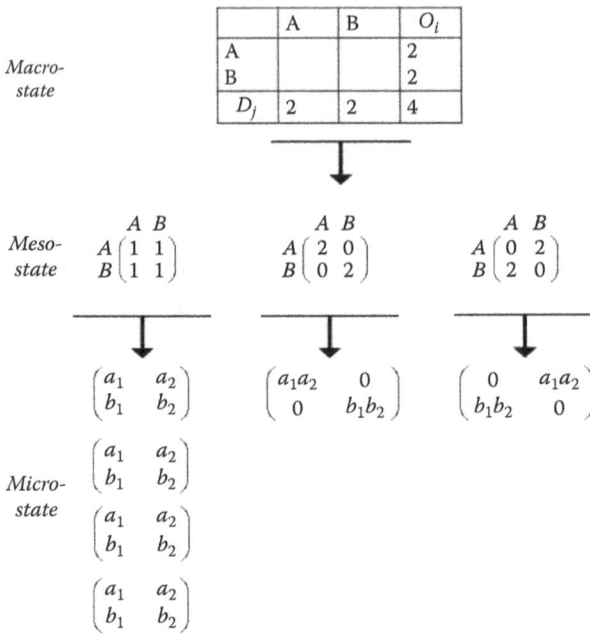

FIGURE 5.2 Micro-states contained within the meso-states and the macro-states.

5.3 SIMPLY CONSTRAINED MODELS

With the simply constrained models, the models constrained by the origin where only information about the origin of journey flows is available can be distinguished from the models constrained by destination where only information about the destinations of the journey flows is known.

5.3.1 MODELS SINGLY CONSTRAINED TO ORIGINS

In the case of the models that are only constrained to the origin, the only available information is about the origin of the flows. Therefore, only constraint Equation 5.3 is considered with these models, and the D_j term has to be substituted by a W_j factor, which is an indicator of the attractiveness of the destination zone for each journey. The model is written as follows:

$$T_{ij} = O_i A_i W_j \exp(-\beta c_{ij}^k) \tag{5.13}$$

where the A_i factor guarantees the fulfilment of the origin constraint:

$$A_i = \left[\sum_j W_j \exp(-\beta c_{ij}^k) \right]^{-1} \tag{5.14}$$

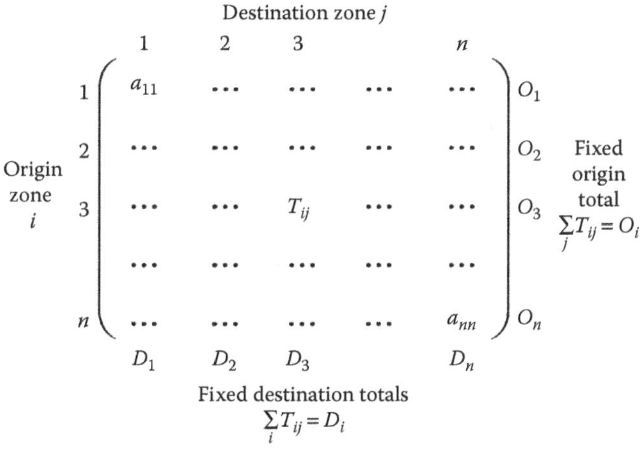

Destination zone j

FIGURE 5.3 Trip matrix of a model singly constrained to origin.

Figure 5.3 shows the matrix resulting from a spatial interaction model singly constrained to origin.

The models constrained only to origin can be interpreted as location models for employment and economic activities as the destinations of the journeys depend on the attractiveness factor W_i of each zone. Further examples are provided in Chapter 9 of models that are applied to the simulation of economic activities location.

5.3.2 MODELS SINGLY CONSTRAINED TO DESTINATIONS

Analogously to the models constrained to origin are the models constrained to the destination. These models use the available information about the destinations of transport flows, so the constraint expressed in Formula 5.4 is fulfilled. In this case, the term O_i has to be substituted by the term W_i, which is an indicator of the trip production from the origins. The result is

$$T_{ij} = W_i D_j B_j \exp(-\beta c_{ij}^k) \tag{5.15}$$

where:

$$B_j = \left[\sum_i W_i \exp(-\beta c_{ij}^k) \right]^{-1} \tag{5.16}$$

Figure 5.4 shows an example of the matrix of a spatial interaction model that is singly constrained to the destination.

The models constrained to destinations have been used to develop residential location models in which the W_i factor is a measure of the area's attraction for residential location. Alternative population and residential location models can be found in Chapter 9.

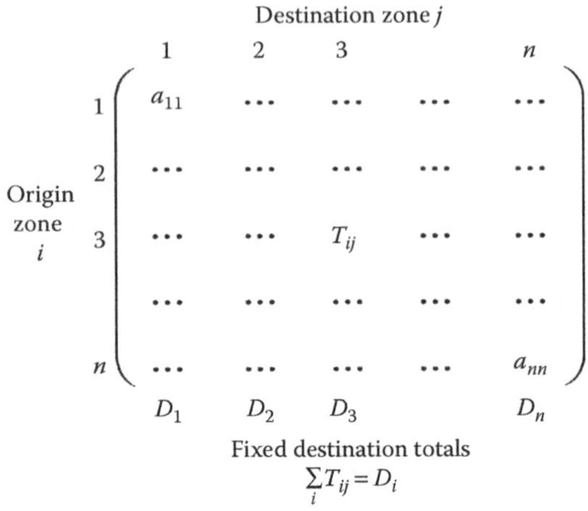

FIGURE 5.4 Trip matrix of a model singly constrained to destination.

5.4 DOUBLY CONSTRAINED SPATIAL INTERACTION MODELS

Doubly constrained models apply where information is available about the level of activity generated in each trip origin zone as well as in each trip destination zone.

These are the kinds of models described in Equations 5.10 through 5.12, which also consider the constraints described in Equations 5.3 and 5.4. As explained earlier, the balancing factors A_i and B_j are obtained by performing an iterative process. Initially, in Equation 5.7, the value of B_j is made equal to 1 and the value of A_i is calculated by substituting this value into Equation 5.8 to obtain the new values of B_j, until the process converges. Figure 5.5 shows the resulting matrix of the doubly constrained spatial interaction model, which, as mentioned previously, contains information about both the origins O_i and the destinations D_j.

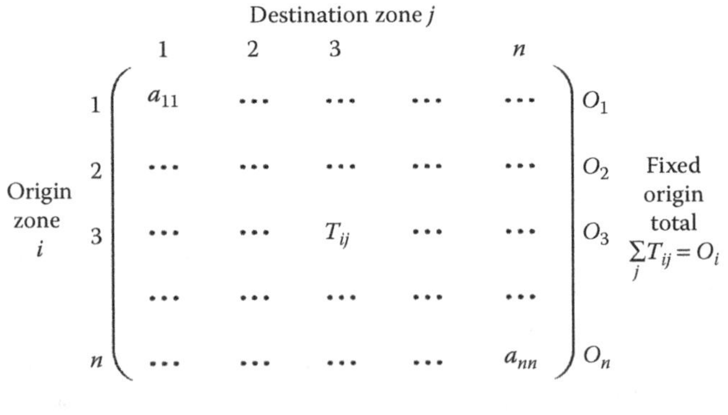

FIGURE 5.5 Trip matrix of a doubly constrained model.

5.4.1 INTERACTION MODELS CONSIDERING TRANSPORT MODE

Among the doubly constrained models, Wilson (1970) proposed considering a combined form of trip distribution and modal choice. This model is based on the previously explained concept of maximum entropy. The aim of the model is to simulate the number of journeys made from origin zones i to the destination zones j using the transport mode k for the population type n. The types of population represent, for example, the ownership or not of a car ($n = 1$, with car ownership and $n = 2$, without car ownership).

Once the initial characteristics of the model have been defined, the following values need to be established:

O_i^n is the number of trips generated in each zone i by each population type n
D_j is the number of attractions in each zone j
c_{ij}^k is the journey cost matrices for each mode k

Therefore, applying all the above we have

$$T_{ij}^{kn} = O_i^n D_j \exp(-\beta c_{ij}^k) A_i^n B_j \tag{5.17}$$

where A_i^n and B_j:

$$A_i^n = \left[\sum_j \sum_{k \in n} B_j D_j \exp(-\beta c_{ij}^k) \right]^{-1} \tag{5.18}$$

$$B_j = \left[\sum_i \sum_n \sum_{k \in n} A_i^n O_i^n \exp(-\beta c_{ij}^k) \right]^{-1} \tag{5.19}$$

These expressions indicate that the sum is found for all the available transport modes k for population type n.

5.5 LAND USE–TRANSPORT INTERACTION MODELS AND SPATIAL INTERACTION

The ideas drawn from spatial interaction theory have been applied through the singly constrained models to simulate the location of population and economic activities. This kind of model starts from the premise that it is more probable for population to locate with the improvement of some zonal accessibility indicator. Accessibility may be measured through an indicator such as

$$V_i = \sum_j M_j \exp(-\beta c_{ij}) \tag{5.20}$$

where M_j represents an attractiveness indicator for zone j, for example, the number of jobs. This indicator combined with the availability of land in each zone was developed by Hansen (1959) to create a residential location model using the following formulation:

$$dR_i = dR \frac{L_i V_i}{\sum_i L_i V_i}$$ (5.21)

where:
  dR is the increase in total population
  dR_i is the increase in population in zone i
  L_i is the availability of land in zone i

These ideas were amplified in the model developed by Lowry (1964). In this model, a compound urban system is defined by three groups of activities: (1) a basic labour sector, (2) a service sector and (3) a residential sector. The goal of the Lowry model is to estimate the location of residents R_i and of jobs in the service sector ^sE_i. Both will, in turn, be influenced by the location of basic sector employment, ^bE_i, which is exogenous to the model and needs to be introduced as an input data. Other necessary input data are the zoning of the study area, the availability of residential land in each zone and a matrix of transport costs C_{ij}. The classification of employment between that which belongs to the basic sector and therefore aimed at the external demand to the study area and that which belongs to the service sector aimed at the internal demand has since been criticised by various authors for its excessive simplicity (Lewis 1976).

The main contribution of the Lowry model is that it manages to relate various sub-models within a complete iterative system and allows the simulation of the economic-spatial structure of an urban system. The diagram in Figure 5.6 shows the functioning of the sub-models, which make up the Lowry model (Barra 1989).

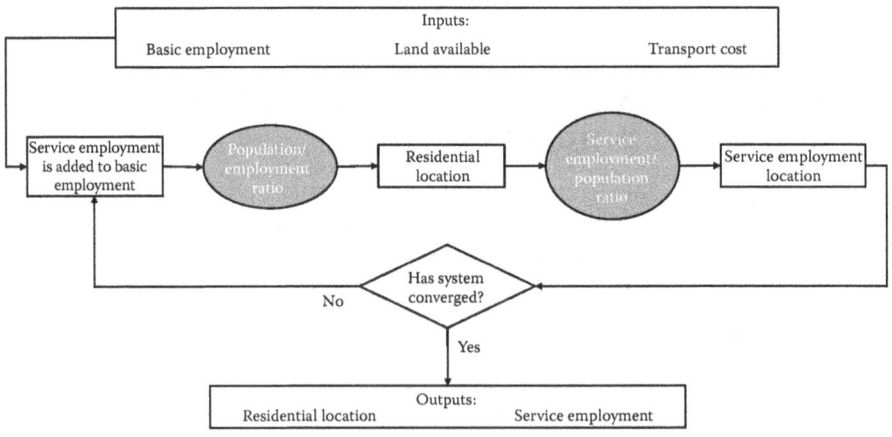

FIGURE 5.6   Structure of the Lowry model.

This method is based on the development of a series of spatial interaction models constrained to origin. The steps required to run the model are presented in the following:

1. Add the employment assigned in the previous iteration to the employment in the basic sector in each zone (this will be zero at the first iteration), where $^*E_i$ is the total employment:

$$^*E_i = {}^bE_i + {}^sE_i \tag{5.22}$$

At each iteration, the new jobs located in the service sector are added to the basic employment in the zone and to the service sector jobs that have been added at the previous iteration.

2. Location of the residents in each zone j (R_{ij}) who work in zone i:

$$R_{ij} = {}^*E_i u A_i L_j \exp(-\beta^r c_{ij}) \tag{5.23}$$

where:

$$A_i = \left[\sum_j L_j \exp(-\beta^r c_{ij}) \right]^{-1} \tag{5.24}$$

and u is the ratio between population and jobs. A_i protects the constraint on the number of residents once it has been multiplied by $^*E_i$ and u. The parameter β^r to be calibrated regulates the effect of the transport costs on the distribution of residents. High values of β^r indicate that the population will locate close to the zone where they work as there is a high cost of travelling.

3. Assignment of service sector jobs to zones j taking into account the number of residents in zones i:

$$^sE_{ij} = R_i v A_i (W_j)^\alpha \exp(-\beta^s c_{ij}) \tag{5.25}$$

where:

$$A_i = \left[\sum_j (W_j)^\alpha \exp(-\beta^s c_{ij}) \right]^{-1} \tag{5.26}$$

where v represents the ratio between the population and jobs in the service sector, W_j represents a term of attractiveness for the service sector jobs and α represents the force of the attraction and thus the tendency of the economic activities to accrete. The parameter β^s must once again be calibrated as it represents the residents' cost to travel to the service sector activities.

After the three steps described earlier, a new iterative process is produced in which a certain number of residents and service sector jobs are relocated at each iteration. This value gets ever smaller until the process converges and a previously defined stop criterion is reached between the iterations.

The Lowry model was later reformed by Garin (1966) who provided a matrix formulation, which produces a direct solution without the need to perform iterations. Echenique (1971) further added a zonal scale land supply sub-model by dividing the initial Lowry model into two blocks: an initial part representing land location and a second part for the location of activities whereby they managed to simulate both the supply and the demand for location.

The Lowry model was the first operational LUTI model based on the spatial interaction theory. This model represents the starting point for the later development of various operational models such as ITLUP/METROPILUS (Putman 1983, Putman and Chan 2001) and LILT (Mackett 1983). Other paradigms used to simulate the location of population and activities in the urban space are addressed in Chapter 9, especially those models based on the random utility theory, currently the ones that are most commonly being used.

REFERENCES

Barra, T. 1989. *Integrated Land Use and Transport Modelling: Decision Chains and Hierarchies, Cambridge Urban and Architectural Studies*. Cambridge, UK: Cambridge University Press.

Batty, M. 1976. *Urban Modelling: Algorithms Calibrations, Predictions, Cambridge Urban and Architectural Studies 3*. Cambridge, UK: Cambridge University Press.

Casey, H. J. 1955. The law of retail gravitation applied to traffic engineering. *Traffic Quarterly* 9 (3):313–321.

Clark, C. 1951. Urban population densities. *Journal of the Royal Statistical Society. Series A (General)* 114 (4):490–496.

Converse, P. D. 1949. New laws of retail gravitation. *Journal of Marketing* 14 (3):379–384.

Echenique, M. 1971. *Urban Systems: Towards an Explorative Model*. London, UK: Centre for Environmental Studies.

Garin, R. A. 1966. A matrix formulation of the lowry model for intrametropolitan activity allocation. *Journal of the American Institute of Planners* 32 (6):361–364. doi:10.1080/01944366608978511.

Hansen, W. G. 1959. How accessibility shapes land use. *Journal of the American Institute of Planners* 25 (2):73–76.

Hoyt, H. 1939. *The Structure and Growth of Residential Neighbourhoods in American Cities*. Washington, DC: Federal Housing Administration.

Huff, D. L. 1963. A probabilistic analysis of shopping center trade areas. *Land Economics* 39 (1):81–90. doi:10.2307/3144521.

Isard, W. 1956. *Location and Space-Economy*. Cambridge, MA: MIT Press.

Lewis, W. C. 1976. Export base theory and multiplier estimation: A critique. *The Annals of Regional Science* 10 (2):58–70. doi:10.1007/bf01303243.

Lowry, I. S. 1964. *A Model of Metropolis, Memorandum*. Santa Monica, CA: Rand Corporation.

Mackett, R. L. 1983. *Leeds Integrated Land-Use Transport Model (LILT)*. Report to the Transport and Road Research Laboratory. Crowthorne, UK: Special Research Branch, Safety and Transportation Department.

Ortúzar, J. D. and Willumsen, L. G. 2011. *Modelling Transport*. Hoboken, NJ: John Wiley & Sons.

Putman, S. H. 1983. *Integrated Urban Models*. 2 vols, Vol. 1: *Research in Planning and Design*. London, UK: Pion.

Putman, S. H. and Chan, S.-L. 2001. The METROPILUS planning support system: Urban models and GIS. In *Planning Support Systems: Integrating Geographic Information Systems, Models, and Visualization Tools*, Brail, R.K. and Klosterman, R.E. (Eds.), pp. 99–128. Redlands, CA: ESRI Press.

Reilly, W. J. 1931. *The Law of Retail Gravitation*. New York: W.J. Reilly.

Rogers, A. 1968. *Matrix Analysis of Interregional Population Growth and Distribution*. Berkeley, UK: University of California Press.

Stewart, J. Q. 1948. Demographic gravitation: Evidence and applications. *Sociometry* 11 (1/2):31–58. doi:10.2307/2785468.

Wilson, A. G. 1970. *Entropy in Urban and Regional Modelling, Monographs in Spatial and Environmental Systems Analysis 1*. London, UK: Pion.

Wilson, A. G. 1974. *Urban and Regional Models in Geography and Planning*. London, UK: Wiley.

Zipf, G. K. 1949. *Human Behavior and the Principle of Least Effort*. Cambridge, MA: Addison-Wesley.

6 Random Utility Theory and Land Use–Transport Interaction Models

Luigi dell'Olio, Rubén Cordera and Ángel Ibeas

CONTENTS

The models based on random utility theory are widely used in various fields of research to solve a multitude of problems. Their more well-known application in transport engineering is the modal choice models, given that these models allow user behaviour to be simulated when choosing the mode of transport for their journey. Random utility models have also been used to simulate journey frequency, route choice and even the choice of destination.

They have also been frequently used with land use–transport interaction (LUTI) models as activity location models or residential location models (see Chapter 9) as well as for choices relating to the transport system. On account of the context of choice with these models coinciding with the zoning of the study area, it highlights the importance of the zoning phase as it has a key effect on the final result.

These models are of a behavioural nature that makes them particularly useful in studying user behaviour when making one or more choice decisions. They can be estimated using aggregated data or individual data. The classic form that was widely used for estimating models using aggregated data, required generic parameters and to avoid the use of specific constants, which in these models expressed the intrinsic characteristics of the different zones within the study area. However, the use of individual data enriches the models because it allows us to study the possible systematic and random variations in the user taste and understand their behaviour much more

69

clearly. Furthermore, specific constants have commonly been omitted in location models because they do not use the complete choice group but rather a sample of alternatives, or because their inclusion was considered to remove generality from the model and make it less sensitive to policy simulation (Waddell 2010). This praxis was affected by an even more serious problem, because, apart from negatively influencing the model's fit, excluding the specific constants did not allow us to explain the spatial characteristics of the different zones, which represents the foundations of LUTI models.

This chapter aims to explain the theoretical foundations behind the models based on the random utility theory. Some of their properties will be studied to understand how to use them appropriately as sub-models within a wider LUTI model. The more basic models will be presented first by progressing to the more advanced models and including an explanation about how the models can be used to consider the spatial dependencies, which frequently appear in transport and land use data.

6.1   RANDOM UTILITY MODELS

The random utility models assume the most frequently used theoretical paradigm to represent the choices made between the discrete alternatives. These models are based on two basic suppositions:

1. Each individual belongs to a homogenous class of individuals, from a behavioural point of view.
2. Each individual makes their choices in a rational way *by maximising their own utility* through the different choice alternatives.

These hypotheses assume the following characteristics:

1. The individuals n belong to a certain homogenous population N. These individuals behave rationally and have perfect information (a choice that maximises their net personal utility).

   We are addressing «homo economicus» subject to environmental restrictions, which could be legal, social, physical or budgetary (in time and money).
2. A generic individual n who is making a choice, considers A_j available alternatives (alternative 1, alternative 2... etc.), which make up their choice group A. The choice group A can be different for different individuals, in other words, there is a certain group $A = (m_1..., m_i..., m_N)$ of available alternatives meaning that the group, which is available to a particular individual n, is $A(n) \in A$.
3. *Each alternative A_j has an associated utility function U_{jn} for an individual n*. The individual n associates a perceived utility function U_{jn}, to each alternative A_j from their group of possible choices and they choose the one that provides them with the most utility, or, they make their choice based on the *maximisation* of their utility.

The *utility* associated to each choice alternative depends on a series of *measurable characteristics or attributes* V_{jn} (rent, transport costs, available housing, etc.), belonging to the individual alternative, $U_{jn} = U_n (V_{jn})$ where V_{jn} is the *vector of the attributes* relative to the alternative j and the individual n making the decision. In other words, *the user chooses an alternative that is based on the attributes belonging to that alternative comparing them with those of the other available alternatives.*

The modeller, as an observer of the system, *does not* possess complete information about all the elements that each individual considers when making their choice and, therefore, assumes that U_{jn} can hold two components:

A measurable part, systematic or *representative*, which is a function of the measured attributes V and a *random portion* that reflects the idiosyncrasy and particular tastes of each individual, as well as measuring any observation errors made by the modeller.

$$U_{jn} = V_{jn} + \varepsilon_{jn} \tag{6.1}$$

$$V_{jn} = \sum_{k=1}^{K} \theta_{jk} X_{jkn} \tag{6.2}$$

The random portion represents the unobserved part of the utility and is normally assumed to have a type-I extreme value distribution (EVI also known as Gumbel) independently and identically distributed or a normal distribution. The way this unobserved part is used has an influence on the creation of different typologies of random utility models. This chapter will not address the Probit type of random utility models because they are not usually applied in the LUTI field as they require a large choice set, which makes them computationally complex as compared to models assuming an EVI-type error distribution.

4. The individual n chooses the alternative, which provides them with the maximum utility; therefore, between alternatives A_j and A_i they will choose alternative A_j if

$$U_{jn} \geq U_{in} \, \forall \, A_j \in A(n) \tag{6.3}$$

or:

$$V_{jn} - V_{in} \geq \varepsilon_{in} - \varepsilon_{jn} \tag{6.4}$$

As the value of $\varepsilon_{in} - \varepsilon_{jn}$ is unknown, this inequality will remain undefined. This is why we can only say that the probability of choosing alternative A_j will be

$$P_{jn} = \text{Prob} \{ \varepsilon_{in} \leq \varepsilon_{jn} + (V_{jn} - V_{in}), \forall \, A_i \in A(n) \} \tag{6.5}$$

If no distribution of probability is specified for the random residuals ε, it is impossible to obtain an analytical expression for the model. However, if we suppose that ε has a mean of zero and therefore the utility function U has a mean of V, the previous equation could be written in the following way:

$$P_{jn} = \int f(\varepsilon)d(\varepsilon) \tag{6.6}$$

In this way, we are able to obtain an analytical expression of the said probability depending on the distribution chosen for the random residuals.

Given a study area, which we are going to divide into A_j zones (see the example in Figure 6.1 for a case with eight zones), the utility of going to live, for example, in zone 3, is given by the following function for each variable X_{3kn} for each user n:

$$U_{3n} = V_{3n} + \varepsilon_{3n} \tag{6.7}$$

$$V_{3n} = \sum_{k=1}^{K} \theta_{3k} X_{3kn} \tag{6.8}$$

The choice of this zone supposes that the net utility perceived by user n is greater than the net utility of going to live in any other zone. This means that it is not the alternative in itself that produces the utility, rather it is the modeller that derives this utility from the characteristics of the alternative to justify the choice made by user n (Lancaster 1966).

Independently of the variables considered in the model, which in many cases depend on the data available to the modeller, an important factor to consider in these models in which a zonal choice is being made is the problem of spatial correlation

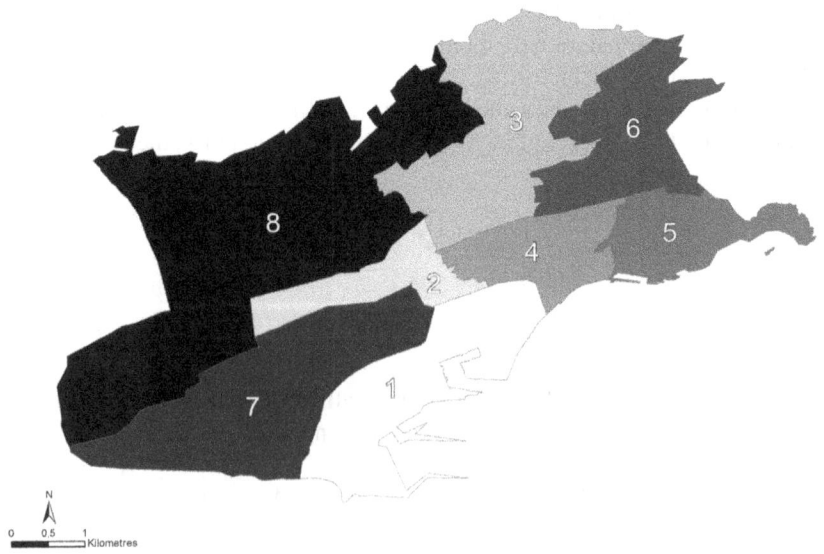

FIGURE 6.1   Example of zoning in the city of Santander.

between the random residuals, above all the residuals associated with nearby zones. This problem has been around for many years, even up to the present day, because it has neither been explored enough nor received the attention it deserves.

6.2 THE MULTINOMIAL LOGIT MODEL

The multinomial logit model (MNL) is the simplest of the discrete choice models and the most popular model that is currently being used. The advantage of this model is that it can obtain the probability P_{jn} of choosing an alternative A_j in a closed way. This is because the model can be generated whilst accepting that the random residuals distribute Gumbel and that the said residuals are independent among all the alternatives and, furthermore, they have the same variance (random residuals independent and identically distributed [IID]) (Domencich and McFadden 1975), so

$$P_{jn} = \frac{\exp(\lambda_j V_{jn})}{\sum_{j=1}^{J} \exp(\lambda_j V_{jn})} \tag{6.9}$$

in which the utility functions generally take a linear form in the parameters and besides, the parameter λ (whose value cannot be estimated separately from the θ_{jk} parameters) is tied to the common variance of the Gumbel variable through the following relationship:

$$\lambda_j^2 = \frac{\pi^2}{6\sigma_j^2} \tag{6.10}$$

The variance and covariance matrix associated to an MNL is a diagonal matrix in which all the covariance is null due to the IID hypothesis.

$$\sum_{\varepsilon} = \sigma^2 \begin{bmatrix} 1 & \cdots & 0 \\ \vdots & \ddots & \vdots \\ 0 & \cdots & 1 \end{bmatrix} \tag{6.11}$$

This supposes a tree structure of the kind shown in Figure 6.2.

6.2.1 SOME PROPERTIES OF MNL

The most important property of the MNL model is the axiom of *independence of irrelevant alternatives* (IIA), which could be enunciated as follows:

> In the case where any pair of alternatives that do not have a null probability of being chosen, the ratio between the probability of both is not influenced by the presence or absence of other additional alternatives, which are present in the choice set.

Luce and Suppes (1965)

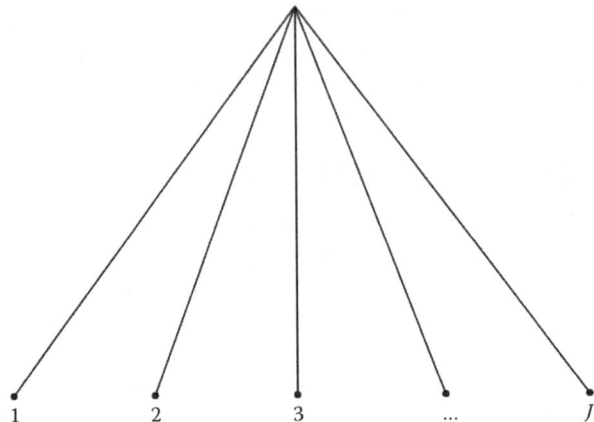

FIGURE 6.2   Tree structure of a multinomial logit model.

In the case of MNL, the quotient: $P_j/P_{j'} = \exp[(V_j - V_{j'})]$ has to be constant and independent of the utility of the other alternatives. This property was initially considered to be an advantage of the model because it allowed the problem of a new alternative to be easily addressed; in other words, the modeller could find the prediction of the market proportion of a new alternative, which was not present when the model was calibrated (if its attributes are known) without having to recalibrate the model. However, this property is currently being perceived as a disadvantage, because it does not allow the model to consider the presence of correlations between alternatives, meaning it could lead to biased predictions. With spatial location models, such as residential or economic activity location models, this property continues to be ignored in many applications even though the presence of correlation between the nearby zones is quite probable. For an MNL to be successfully used in these cases, the modeller needs to be sure that no correlation between the alternatives is present, which can only be checked by estimating more complex models that consider these relationships as well as conducting a statistical test to discard the hypothesis.

Furthermore, when there are a high number of alternatives, as is the case with location and destination choice, the researcher can demonstrate that unbiased parameters are obtained by estimating the model with only a *random* sample taken from the set of available choices (e.g. seven destinations for each user) (McFadden 1977). The models that do not possess such a property may require a great deal of computing time for more than 50 alternatives, even where their estimation is not overly complicated. Unfortunately, in the contexts of destination choice or location choice, it is not difficult to surpass this figure in average-sized zoning systems (Ortúzar and Willumsen 2011).

6.2.2   Elasticities

Elasticity can be defined as an adimensional measurement, which indicates the relationship between the percentage change of a variable and the percentage change in the demand, ceteris paribus.

With discrete choice models, where in addition to a given alternative, we could have competing alternatives, we have to differentiate between the direct elasticity and the cross-elasticity.

The definition provided earlier is valid for direct elasticities, but the cross-elasticity is defined by Louviere et al. (2000) as the percentage change in the probability of choosing an alternative with respect to the percentage change of an attribute in a competing alternative.

The direct point elasticity for the MNL is given by

$$E_{X_{jkn}}^{P_{jn}} = \frac{\partial P_{jn}}{\partial X_{jkn}} \frac{X_{jkn}}{P_{jn}} \qquad (6.12)$$

which, according to Louviere et al. (2000) represents the elasticity of the probability of an alternative j for user n with respect to a marginal change in attribute k for alternative i.

This formula may be expressed in its more simplified way as follows:

$$E_{X_{jkn}}^{P_{jn}} = -\theta_{jk} X_{jkn}(1 - P_{jn}) \qquad (6.13)$$

And the cross-elasticity as

$$E_{X_{ikn}}^{P_{jn}} = -\theta_{ik} X_{ikn} P_{in} \qquad (6.14)$$

Knowledge about elasticities is important, because it uses an adimensional indicator to help us understand the importance of a variable and particularly its effect on the demanded quantity.

6.3   THE NESTED LOGIT MODEL

As seen in the previous section, the structure of the MNL covariance matrix may present problems when the alternatives are not independent, that is, when groups of alternatives that are more similar than others are present. For example, this could occur when choosing between a public transport and a private car or in the choice between zones having similar characteristics in spatial location models.

The probit model, which could be derived from a multivariate normal distribution (instead of a Gumbel IID), presents a completely general covariance matrix allowing the researcher to address the previous problem. However, it is not easy to solve this model except in cases where there are very few (up to 3) alternatives (Bouthelier and Daganzo 1979).

An alternative to the MNL and the probit is nested logit (NL), which can specify a structure of correlation overcoming the problems derived from the IIA axiom and without the unnecessary complications involved with the probit model.

The NL models can be estimated sequentially or simultaneously and allow the modeller to address only as many interdependencies between alternatives as nests have been specified in the structure; furthermore, the alternatives of one nest cannot be correlated with those of other nests. This crossed correlation effect may need to be tested in a context of modal choice between the mixed modes using more general methods such as those provided by the probit models.

The search for the best nested structure sometimes requires the tentative analysis of multiple structures. It is easy to see *a priori* that the number of possible groupings increases geometrically with the number of alternatives (Sobel 1979). Although the existence of information *a priori* together with the modeller experience may be of some assistance in this task, the search could take much longer than with the simpler MNL.

The sequential calibration of the model is simple and made possible by the availability of MNL software; furthermore, it produces consistent estimators (in fact, as the amount of data increases, these estimations converge towards the true values of the parameters). In spite of this, the approach does present various potential problems; for example, if insufficient data are available to estimate lower level models, the estimations may be inefficient, both because information has been omitted on the lower level and because the errors they cause may be transmitted to the upper levels. It has been empirically demonstrated that some interesting structures cannot be tested or, even worse, that the method may lead to the rejection of structures that are demonstrably better (Hensher 1986, Ortuzar et al. 1987).

If λ_b represents the scaling parameter on the highest level of the tree and $\mu_{j/b}$ is the scaling factor at the base of the tree, the utility of the generic alternative at the lower level of the tree, belonging to nest b will be

$$U_j = \mu_{j/b} \sum_{k=1}^{K} \theta_k X_{jk} + \varepsilon_j \tag{6.15}$$

where:

$$\mu_{j/b} = \frac{\pi^2}{6\mathrm{var}(\varepsilon_{j/b})}$$

Therefore, if the variance of the error increases, $\mu_{j/b}$ decreases, and if the variance falls, $\mu_{j/b}$ increases at the same time as the observed utility component also increases.

The utility on the upper level of the tree's structure is linked to the utility of the alternatives contained in the nest through

$$\lambda_b \left(\frac{1}{\mu_{j/b}} \log \left[\sum_{b \in j} \exp(\mu_{j/b} V_{j/b}) \right] \right) \tag{6.16}$$

which represents the Logsum (Cascetta 2009, Ortúzar et al. 2014).

Normally, for the model to be identifiable, we need to normalise by making $\mu_{j/b} = 1$ (RU1) or $\lambda_b = 1$ (RU2) (Carrasco and Ortúzar 2002, Hensher and Greene 2002, Hensher et al. 2015). Knowing that NL introduces correlation between the utilities of the alternatives contained in the same nest (Ben-Akiva and Lerman 1985), the covariance for the alternatives j and j' included in the same nest b is given by the equation:

$$\text{Cov}\left(U_{(j/b)}, U_{(j'/b)}\right) = 1 - \left(\frac{\lambda_b}{\lambda_{(j/b)}}\right) \tag{6.17}$$

Therefore, considering the tree structure, the probability of choosing alternative j is given by

$$P_j = P_{j/b} P_b = \frac{\exp\left(\mu_{j/b} V_{j/b}\right)}{\sum_{i \in J_b} \exp\left(\mu_{i/b} V_{i/b}\right)}$$

$$\frac{\exp\left[(\lambda_b / \mu_{j/b}) \log\left(\sum_{b \in J_b} \exp(\mu_{j/b} V_{j/b})\right)\right]}{\sum_{b=1}^{B} \exp\left[(\lambda_b / \mu_{j/b}) \log\left(\sum_{i \in J_b} \exp(\mu_{i/b} V_{i/b})\right)\right]} \tag{6.18}$$

where $P_{j/b}$ is the conditional probability of choosing alternative j, which belongs to nest b, and P_b is the marginal probability of choosing nest b.

If we consider the example shown in Figure 6.1 in which 8 zones represent the city of Santander (Spain) and in a residential choice context, we suppose that zones 3, 4, 5 and 6 are correlated among themselves, as are zones 7 and 8 on the other side. In this case, we have a structure of the variance–covariance matrix Σ_ε of the following kind:

$$\Sigma_\varepsilon = \begin{pmatrix} \sigma^2 & 0 & 0 & 0 & 0 & 0 & 0 & 0 \\ 0 & \sigma^2 & 0 & 0 & 0 & 0 & 0 & 0 \\ 0 & 0 & \sigma^2 & \sigma_{b'} & \sigma_{b'} & \sigma_{b'} & 0 & 0 \\ 0 & 0 & \sigma_{b'} & \sigma^2 & \sigma_{b'} & \sigma_{b'} & 0 & 0 \\ 0 & 0 & \sigma_{b'} & \sigma_{b'} & \sigma^2 & \sigma_{b'} & 0 & 0 \\ 0 & 0 & \sigma_{b'} & \sigma_{b'} & \sigma_{b'} & \sigma^2 & 0 & 0 \\ 0 & 0 & 0 & 0 & 0 & 0 & \sigma^2 & \sigma_{b''} \\ 0 & 0 & 0 & 0 & 0 & 0 & \sigma_{b''} & \sigma^2 \end{pmatrix} \tag{6.19}$$

With a tree structure as reflected in Figure 6.3.

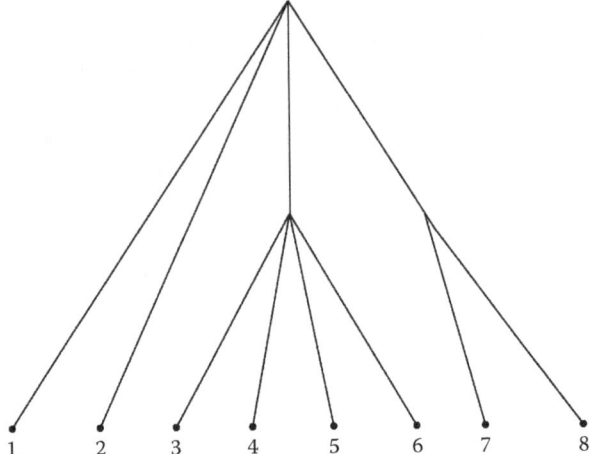

FIGURE 6.3 Structure of a nested model considering correlation between the errors in two groups of alternatives.

6.4 THE CROSS-NESTED LOGIT MODEL

The cross-nested logit (CNL) model appeared as a generalisation of NL, which considered that an alternative may belong to more than one nest with different degrees of inclusion. Starting from the formulation of the NL model, the probability of choosing generic alternative j is given by

$$P_j = \sum_{b \in J_b} P_{j/b} P_b \qquad (6.20)$$

where b is the generic node of a single-level nested structure.

The degree that alternative j belongs to nest b is now going to be denominated as α_{jb} and it has to fulfil the following conditions:

$$0 \le \alpha_{jb} \le 1 \qquad (6.21)$$

$$\sum_{b \in J_b} \alpha_{jb} = 1 \quad \forall \, j \qquad (6.22)$$

The analytical expression of $P_{j/b}$ being as follows:

$$P_{j/b} = \frac{\alpha_{jb}^{1/\delta_b} \exp\left(\mu_{j/b} V_{j/b}\right)}{\sum_{i \in J_b} \alpha_{ib}^{1/\delta_b} \exp\left(\mu_{i/b} V_{i/b}\right)} \qquad (6.23)$$

In addition, the expression of P_b is as follows:

$$P_b = \frac{\left[\sum_{i \in J_b} \alpha_{ib}^{1/\delta_b} \exp\left(\mu_{i/b} V_{i/b}\right)\right]^{\delta_b}}{\sum_{b'} \left[\sum_{i \in J_{b'}} \alpha_{ib'}^{1/\delta_{b'}} \exp\left(\mu_{i/b} V_{i/b'}\right)\right]^{\delta_{b'}}} \qquad (6.24)$$

where J_b is the group of alternatives belonging to nest b, $\mu_{j/b}$ is the parameter associated with a lower level nest and λ_b is the parameter associated with the upper level and δ_b is the ratio between the latter two. Combining Equations 6.23 and 6.24 leads to

$$P_j = \sum_{b \in J_b} P_{j/b} P_b = \sum_{b \in J_b} \left(\frac{\alpha_{jb}^{1/\delta_b} \exp\left(\mu_{j/b} V_{j/b}\right)}{\sum_{i \in J_b} \alpha_{ib}^{1/\delta_b} \exp\left(\mu_{i/b} V_{i/b}\right)} \cdot \frac{\left[\sum_{i \in J_b} \alpha_{ib}^{1/\delta_b} \exp\left(\mu_{i/b} V_{i/b}\right)\right]^{\delta_b}}{\sum_{b'} \left[\sum_{i \in J_{b'}} \alpha_{ib'}^{1/\delta_{b'}} \exp\left(\mu_{i/b'} V_{i/b'}\right)\right]^{\delta_{b'}}} \right)$$

$$= \frac{\sum_{b \in J_b} \left[\alpha_{jb}^{1/\delta_b} \exp\left(\mu_{j/b} V_{j/b}\right) \cdot \left[\sum_{i \in J_b} \alpha_{ib}^{1/\delta_b} \exp\left(\mu_{i/b} V_{i/b}\right)\right]^{\delta_b - 1}\right]}{\sum_{b \in J_b} \left[\sum_{i \in J_b} \alpha_{ib}^{1/\delta_b} \exp\left(\mu_{i/b} V_{i/b}\right)\right]^{\delta_b}}$$

$$(6.25)$$

If $\lambda_b = \mu_{j/b}$, then $\delta_b = 1$; it is easy to demonstrate that the multinomial logit is obtained. Therefore, δ_b represents the correlation between the different alternatives. The covariance and variance are expressed in the following way:

$$\mathrm{Cov}(\varepsilon_i, \varepsilon_j) = \frac{\pi^2 \lambda_b^{\,2}}{6} \sum_{b \in J_b} (\alpha_{ib})^{1/2} (\alpha_{jb})^{1/2} (1 - \delta_b^{\,2}) \qquad (6.26)$$

$$\mathrm{var}\left(\varepsilon_i\right) = \frac{\pi^2 \lambda_b^{\,2}}{6} \sum_{b \in J_b} (\alpha_{ib})^{1/2} (\alpha_{ib})^{1/2} = \frac{\pi^2 \lambda_b^{\,2}}{6} \sum_{b \in J_b} (\alpha_{ib}) = \frac{\pi^2 \lambda_b^{\,2}}{6} \qquad (6.27)$$

On account of the flexibility shown by this model when estimating correlation structures it is perhaps one of the better at adapting to modelling the existing correlation between different urban zones when researching residential or economic activities location. Due to the heterogeneity of the zones, which may appear in a study area, this model is able to assume that a certain zone could belong to one or more nests; in other words, it can show correlation in the random term with more than one zone.

In the example of the NL model presented in the previous section, due to their peripheral nature in relation to the urban centre, zones 3 and 6, similarly to zones 7 and 8, could show correlation with each other and with zones 4 and 5. This possible correlation could be specified using CNL with two nests at the same time, as shown by the dotted lines in Figure 6.4.

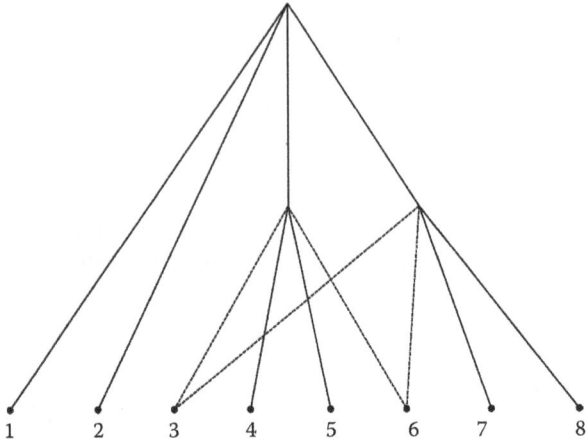

FIGURE 6.4 Structure of a cross-nested logit model considering two alternatives belonging to two nests.

6.5 LOGIT MODELS WITH RANDOM PARAMETERS AND ERROR COMPONENTS

Through the use of NL, we have managed to consider the correlation between the errors of the more similar alternatives. However, this kind of model is unable to consider random variations in the user taste. One way of considering these variations consists of assigning a distribution of probability to the parameters associated with certain variables, thereby providing specific parameters for each user. The ideal situation is that these variations are explained by systematic factors (e.g. systematic variations in the taste obtained by introducing interactions with the socioeconomic variables). The models that fulfil these conditions are the mixed logit (ML) or random parameter logit model. These models assume that some parameters are random and follow a probability distribution, which will be continuous throughout the sampled population.

Therefore, the probability of choosing alternative j is given by the following equation:

$$\text{Prob}\left(\frac{j_n}{x_{jn}z_n v_n}\right) = \frac{\exp(V_{jn})}{\sum_{j=1}^{J} \exp(V_{jn})} \tag{6.28}$$

where:

$$V_{jn} = \theta'_n x_{jn} \tag{6.29}$$

$$\theta_n = \theta + \Delta z_n + \Gamma v_n \tag{6.30}$$

where:

x_{jn} represents the K attributes of alternatives j for user n

z_n is the mean of the M attributes representing the tastes of individual n (the characteristics of individual n)

v_n is the vector associated to the K variables, random with the mean zero, variance known (normally equal to 1) and covariance equal to zero

Based on this formulation, we can conclude that this model includes heterogeneities, which are observed (Δz_n) and not observed (Γv_n) for the parameters representing the variations in taste of users n.

Therefore, Δ is the KxM matrix of parameters associated with the characteristics of the individuals and Γ is the Cholesky matrix (lower triangular matrix).

It is a KxM matrix because the non-random parameters in Γ are assumed to equal zero and vice versa in Δ. When $\Delta = \Gamma = 0$, the classic MNL is obtained. Even so, it is still possible to consider systematic variations in taste using the classic MNL by introducing interactions with, for example, socio-economic variables.

The expected probability can now be obtained considering the random distribution of the parameters, which at this time is not closed:

$$E(P_n^*) = \int_\beta P_n^*(\theta) f\left(\frac{\theta}{\Omega}\right) d\theta \tag{6.31}$$

where $f(\theta/\Omega)$ is the multivariate probability density function of θ given the distributional parameters Ω.

The randomly varying parameters are generally indicated using the following equation:

$$\theta_{nk} = \overline{\theta_k} + \eta_k z_{nk} \tag{6.32}$$

where:

$\overline{\theta_k}$ represents the mean of the distribution

η_k is the deviation or dispersion of the sampled users' preferences around the mean of the marginal utility

z_{nk} are the random values obtained from the specified distribution

Based on this formulation, the preferences vary according to a determined distribution among the n individuals.

We can specify an alternative formulation to the above-mentioned formula in order to reach a heteroscedastic interpretation of the utility. The model, known as the error components model (EC) considers that there may be alternatives to the utility of which have some associated shared error component, distributed with a mean of zero and with a standard deviation to estimate.

Models with error components use a series of dummy variables to group the alternatives into nests. This makes it possible to obtain structures of covariance, which

are more complicated than with the NL. In this case, the utility function can be
specified in the following way:

$$U_{jn} = \sum_{k=1}^{K} \theta_k x_{jnk} \pm \sum_{l=1}^{L} \eta_l z_{ln} d_{lb} + \varepsilon_{jn} \qquad (6.33)$$

where d_{lb} is equal to 1 if j belongs to the nest and is equal to 0 if not.

The special characteristic of the EC model is associated to the alternative and not
to the attributes. Each error component represents the residual variance of the error,
which correlates these two alternatives. Their covariance structure can be expressed
using the following equation if j' belongs to nest b:

$$\text{Cov}(U_{jn'}, U_{jn}) = E(\eta_{j'} z_{jn'} d_{jb'} + \varepsilon_{jn'})'(\eta_j z_{jn} d_{jb} + \varepsilon_{jn}) = \Omega_b \qquad (6.34)$$

This covariance will be equal to zero in any other case.

The variance for each alternative belonging to nest b will be

$$\text{Var}(U_{jn}) = E(\eta_j z_{jn} d_{jb} + \varepsilon_{jn})^2 = \Omega_b + \frac{\pi^2}{6\sigma_n^2} \qquad (6.35)$$

It is easy to imagine that this model can be applied with both fixed parameters and
those considering systematic and/or random variations in taste.

6.6 WAYS OF CONSIDERING SPATIAL DEPENDENCE IN RANDOM UTILITY MODELS

Random utility models are not born as models designed to study cases of spatial
dependence between alternatives. However, they are widely used in LUTI modelling, mainly because of their capacity to study user behaviour, especially in terms
of their location choices. As the study of spatial relationships is one of the basic
characteristics of LUTI models, the simplest utility models, such as MNL, fail
when they consider that the ε_{jn} errors associated with the utility functions of each
choice alternative that are distributed independently and identically. This supposition is not acceptable if we are addressing residential location models or economic
activities location models, given that on the one hand we would be assuming that
the variance of the errors associated with each alternative was identical (some-
thing a problem cannot assume, but needs to make us reflect) and on the other
hand that the covariance is null. This means it is assumed that there is no covariance between the choice alternatives and that, therefore, the alternatives, that could
be zones with very similar economic and socio-demographic characteristics, are
independent of each other.

This problem can be solved by considering a sufficiently aggregated zoning.
However, the zones that are contiguous to each area could still present situations of

correlation. Furthermore, the problem of having a more generic model with scarce predictive capacity would persist.

To solve this problem, Bhat and Guo (2004) proposed a location model with spatial correlation. On similar lines, Ibeas et al. (2013) applied NL and CNL models to consider covariance between some choice alternatives and used the likelihood ratio test to show that both the NL and the CNL models had a better fit than the MNL model with a confidence level of 95% by considering the existence of spatial correlation between the alternatives.

Considering that the NL and CNL models allow the modeller to assess situations of spatial correlation and find better fits, the use of ML and EC models is even better because they allow the study of more sophisticated and flexible error structures. Furthermore, these last two models allow the practitioner to study the user behaviour in greater detail, given that they not only consider systematic variations in taste but also random variations. The only problem, which partially exists with these models is their computational complexity tied to the high number of choice alternatives, available observations and the number of parameters to estimate. However, these problems can be overcome by using a greater computing power.

Even though this battery of resources allows us to consider the relationships between the different zones in a flexible way, other interesting resources are available, which can help us to fine tune our models and more thoroughly understand the spatial relationships that may exist within the data. One of these additional resources consists of using specific constants. Specific constants are a useful tool if we wish to use the model for making predictions, given that a complete group of constants can reproduce the market share of the alternatives. Specific constants also have the advantage of allowing the study area to be clustered without limiting the model from the point of view of considering the presence of covariance in the error distribution that is associated with the choice alternatives.

The model can therefore be forced into presenting equal constants for similar zones by clustering the study area and giving the model the supposition that some of these zones may have similar characteristics and, therefore, the same specific constant. It is clear that this hypothesis should be checked through a procedure such as a step-wise fashion to demonstrate that the constants are significantly similar. If this hypothesis is not checked, the modeller runs the risk of forcing the model by introducing constraints that could lead to a bias in the predictions.

An alternative method of clustering the study area consists of using an accessibility variable. This research is still in the development phase, but Sigismondo (2015) shows that the accessibility normally introduced into residential location models with a specific parameter may have a different weighting in different zones within the study area and could even have no relevance at all in some of the city's zones. This conclusion is very important because it means that the features of the model in terms of fit can be improved leading to more accurate predictions and a better understanding of the spatial dynamics, which will lead to better policies being introduced. In the specific case of the Sigismondo (2015) study on residential location, accessibility was significant in the transition zones between the city centre and the periphery, whereas it had little significance in both the centre and the periphery themselves. This highlights that accessibility is probably not a relevant variable in either the centre or the periphery. Nevertheless, the same is not true of the transition zones, hybrid zones without all the attractions of the central zones or the advantages of the peripheral zones, which need to be accessible for the population to locate there.

Therefore, not only the choice of model is determinant but also the level of aggregation (zoning) and the way it is specified to achieve more exact predictions are determinants.

REFERENCES

Ben-Akiva, M. E. and Lerman, S. R. 1985. *Discrete Choice Analysis: Theory and Application to Travel Demand*. MIT Press Series in Transportation Studies 9. Cambridge, MA: MIT Press.

Bhat, C. R. and Guo, J. 2004. A mixed spatially correlated logit model: Formulation and application to residential choice modeling. *Transportation Research Part B: Methodological* 38 (2):147–168. doi:10.1016/s0191-2615(03)00005-5.

Bouthelier, F. and Daganzo, C. F. 1979. Aggregation with multinomial probit and estimation of disaggregate models with aggregate data: A new methodological approach. *Transportation Research Part B: Methodological* 13 (2):133–146.

Carrasco, J. A. and Ortúzar, J. D. 2002. Review and assessment of the nested logit model. *Transport Reviews* 22 (2):197–218.

Cascetta, E. 2009. *Transportation Systems Analysis: Models and Applications*, 2nd ed. Springer Optimization and its Applications. New York: Springer.

Domencich, T. A. and McFadden, D. 1975. *Urban Travel Demand: A Behavioral Analysis, Contributions to Economic Analysis 93*. Amsterdam, the Netherlands: Elsevier.

Hensher, D. A. 1986. Sequential and full information maximum likelihood estimation of a nested logit model. *The Review of Economics and Statistics* 68 (4):657–667.

Hensher, D. A. and Greene, W. H. 2002. Specification and estimation of the nested logit model: Alternative normalisations. *Transportation Research Part B: Methodological* 36 (1):1–17.

Hensher, D. A., Rose, J. M. and Greene, W. H. 2015. *Applied Choice Analysis*, 2nd ed. Cambridge, UK: Cambridge University Press.

Ibeas, Á., Cordera, R., Olio, L. and Coppola, P. 2013. Modelling the spatial interactions between workplace and residential location. *Transportation Research Part A: Policy and Practice* 49:110–122. doi:10.1016/j.tra.2013.01.008.

Lancaster, K. J. 1966. A new approach to consumer theory. *Journal of Political Economy* 74:132.

Louviere, J. J, Hensher, D. A. and Swait, J. D. 2000. *Stated Choice Methods: Analysis and Applications*. New York: Cambridge University Press.

Luce, R. D. and Suppes, P. 1965. Preference, utility, and subjective probability. In *Handbook of Mathematical Psychology*, Luce, R. D., Bush, R. R. and Galanter, E. (Eds.), pp. 252–410. New York: Wiley.

McFadden, D. L. 1977. Modelling the choice of residential location. Cowles Foundation for Research in Economics, Yale University.

Ortuzar, J. D., Achondo, F. J. and Ivelic, A. M. 1987. Sequential and full information estimation of hierarchical logit models: Some new evidence. *Proceedings 11th Triennial Conference on Operations Research*. Buenos Aires, Argentina, August 1987.

Ortúzar, J. D., Cherchi, E. and Rizzi, L. 2014. Transport research needs. *Handbook of Choice Modelling*. doi:10.4337/9781781003152.00040.

Ortúzar, J. D. and Willumsen, L. G. 2011. *Modelling Transport*. West Sussex, UK: John Wiley & Sons.

Sigismondo, G. 2015. Modeling residential location considering systematic and random variation in tastes. Master Thesis, Politecnico di Bari, Italy.

Sobel, K. L. 1979. Travel demand forecasting by using the nested multinomial logit model. *Transportation Research Record* 775:48–55.

Waddell, P. 2010. Modeling residential location in UrbanSim. In *Residential Location Choice*, Pagliara, F., Preston, J. and Simmonds, D. (Eds.), pp. 165–180. Berlin, Germany: Springer.

7 Optimisation and Land Use–Transport Interaction Models

Sara Ezquerro and Borja Alonso

CONTENTS

7.1 OPTIMISATION MODELS APPLIED TO THE INTERACTION BETWEEN LAND USE AND TRANSPORT

Optimisation models are applied in the field of land use–transport interaction (LUTI) modelling to help solve urban planning problems and make predictions about the location of activities and their spatial interaction. The decisions made about the locations of different land uses and their patterns of development, be they residential, commercial, industrial or other, impact the generation of traffic within the urban areas and the development of public services and installations. Optimisation models are a tool for evaluating and analysing possible alternatives to the optimal development provided by the model as its solution.

For many years, optimisation models have been used by numerous researchers as support tools for decision-making (Markowitz 1956). These kinds of models are formed from three basic elements: (1) the problem's variables, (2) an objective function to be optimised and (3) a group of constraints. Decision variables are a key element in the formulation of optimisation problems and especially in LUTI modelling given the great variety of variables to be considered, such as land development costs, general journey costs, journey times, various social or environmental aspects, accessibility and more. The objective function is the element used to decide the optimal values of the decision variables and represents a quantitative measurement of the system, which needs to be maximised or minimised. In the case of LUTI models, this could mean, among other objectives:

- Minimise the cost of the real estate development
- Minimise the journey to work costs
- Maximise the social benefits
- Minimise the transport costs

The objective function is subjected to a group of constraints, which could include: the maximum and minimum levels of development allowed for each zone for different activities, environmental constraints and controls on the compatibility of activities and so on. Optimisation models try to find the value that the decision variables need to take to optimise the objective function while satisfying the group of specified constraints.

Optimisation techniques applied to LUTI models have evolved over the years and the following sections (Sections 7.2–7.4) will summarise some of the main models that have been proposed in the literature.

7.2 THE HERBERT–STEVENS MODEL

The model developed by Herbert and Stevens (1960) was one of the first models to address the optimisation of land use and estimate the optimal location of households on the residential land. This model was developed as part of a wider model, which was designed to locate all types of land use.

In order to distribute households on residential land, Herbert and Stevens assumed, starting from the residential location theory of Alonso (1964), that households considered the following factors when deciding their location: the overall budget available to them, the products on the market and the cost of those products; in this case, housing and its financial cost. The objective function of this model is the maximisation of the prices paid by the households:

$$\text{Max } Z = \sum_{K=1}^{U} \sum_{i=1}^{n} \sum_{h=1}^{m} X_{ih}^{k} \left(b_{ih} c_{ih}^{k} \right) \tag{7.1}$$

subject to

$$\sum_{i=1}^{n} \sum_{h=1}^{m} S_{ih} X_{ih}^{K} \leq L^{K} \quad \left(K = 1, 2, \ldots, U \right) \tag{7.2}$$

$$\sum_{K=1}^{U} \sum_{h=1}^{m} X_{ih}^{K} = N_i \quad \forall X_{ih}^{K} \geq 0 \quad (K = 1, 2, \ldots, U)(i = 1, 2, \ldots, n)(h = 1, 2, \ldots, m) \tag{7.3}$$

where:

U are the areas coming from the subdivision of the region being studied and are indicated by the superscript $K = 1, 2,...,U$

n are the groups of households, indicated by the superscript $i = 1,2,...,n$

m are the residential bundles, indicated by the superscript $h = 1,2,...,m$

b_{ih} is the residential budget assigned by the household group i for the purchase of residential bundle h

c_{ih}^K is the annual cost for a household from group i of the residential bundle h in area K

s_{ih} is the number of acres in a place used by a household from group i if they use residential bundle h

L^K is the amount of land available for housing in area K at an iteration of the model

N_i is the number of households in group i that have located in region i during a specific iteration

X_{ih}^K is the number of households in group i using residential bundles located by the model in area K

Where the maximum amount that a household can pay in an area for a particular place, is the household's ability to pay for a place in that area, which can be defined as the difference between the available finance and the cost of that place. Constraint 7.2 avoids the assignment of more land use than is available, whereas constraint 7.3 means that all the projected households are located somewhere.

In other words, the model maximises the prices paid by the households to the real estate owners. This income maximisation means that the distribution of the households is Pareto optimal, or, no household can locate in another place without increasing the prices paid by another household and at the same time reduce the prices on an aggregated level.

The model can similarly be expressed as a dual problem in which the optimal solution for the two models is identical. In the dual problem, the subsidies received by the households to obtain housing are minimised, thereby incorporating a variable related to public life. The dual problem substitutes the variable X_{ih}^K of the first problem with the annual rent per unit of land in an area K (r^K) and an annual subsidy per household for all the households in group (v_i), making the objective function:

$$\min Z' = \sum_{K=1}^{U} r^K L^K + \sum_{i=1}^{n} v_i(-N_i) \tag{7.4}$$

s.t.

$$s_{ih} r^K - v_i \geq b_{ih} - c_{ih}^K \quad \forall r^k \geq \quad (K = 1,2,...,U)(i = 1,2,...,n)(h = 1,2,...,m) \tag{7.5}$$

The main drawback of the Herbert–Stevens model is that it makes an efficient land use distribution, which sometimes can be very unrealistic and with a limited role for transport. However, it is the foundation on which later models were developed to simulate the interaction between land use and transport. This model forms part of the first generation LUTI models developed during the 1960s in the United States. They provided limited explanation and evaluation of their results, which prompted the development of new models (Wegener 1994), such as Technique for Optimum Placement of Activities into Zones (TOPAZ) to overcome these limitations.

New studies based on bi-level equilibrium models appeared after the TOPAZ model and its later varieties (Brotchie and Sharpe 1975, Brotchie et al. 1980, Foot 1981). These models were developed to address the influence of two agents such as the planner and the population, who wish to optimise their different but interdependent objectives.

7.3   MULTI-OBJECTIVE OPTIMISATION MODELS

As their name suggests, multi-objective optimisation models optimise various objectives at the same time. Conflicts appear between these objectives because an improvement made to one of them will cause worsening to another; meaning a decision will be required as to the best overall solution to the problem.

7.3.1   Technique for Optimum Placement of Activities into Zones Model

The TOPAZ model was developed in the 1970s to provide the town planner with a group of alternative solutions using mathematical techniques to determine land use patterns in urban areas. TOPAZ can assign future land use classifications to different zones in a study area and is a useful tool for solving urban and regional planning problems, as well as problems associated with the planning of urban services.

TOPAZ is a linear programming model that considers the urban system to be organised by the interaction of different components such as a group of activities, the suitable zones for locating those activities and the routes that interconnect them. The objective function of the TOPAZ model varies depending on the objectives that need to be optimised:

- Minimise the total costs of urban housing, services and transport
- Minimise the consumption of petrol and journeys being made
- Maximise accessibility
- Maximise the suitability of the activities to the zones
- A weighted combination of the earlier, or others

Any of the previous objective functions should be subjected to two basic constraints:

- All the activities are located in some area and none of the areas contains more activities than they are able to support.
- The trips between zones obey certain rules of behaviour.

Further restrictions can also be included such as the prohibition of growth in certain parts of the city, the restriction of certain activities in certain zones or, on the contrary, oblige certain activities to locate in particular zones. The mathematical base of the TOPAZ model has the advantage of being able to adapt to an ample range of situations, which gives the model-added value.

It is important to highlight the flexibility of the TOPAZ model and its ability to perform sensitivity analysis and provide efficient support to the decision-making process during the planning procedure and the design of alternatives. The flexibility to incorporate new objectives and constraints, with relatively little change made to the general mathematical structure of the programme and its ability to indicate the sensitivity of the results to changes in hard to predict data are some of the model's strong points. An example of input data, which is difficult to predict, are the future population levels that are always an estimation with a variable level of error. Furthermore, common interests can have priority over individual interests in the context of the design of urban planning and urban services. The maximum benefit for thousands of personal decisions relating to each individual does not necessarily imply the maximum social benefit. Another of the advantages is the ability of TOPAZ to highlight extreme situations, that is, solutions that are very different from the existing situation and perhaps, at first sight, could turn out to be unrealistic. However, the model can also demonstrate potential for growth in an area, which had not occurred to the planners. It has the ability to obtain the optimum solution as well as those solutions, which are close to it. Finally, TOPAZ is a tool for evaluating alternatives based on multiple criteria such as the costs associated with the public transport, private transport and the environmental considerations and so on.

Nevertheless, the model does have certain limitations; among them its partial usefulness in many planning situations as any solution it generates will be very complicated to completely put into practise. However, these kinds of solutions can probably be useful in supporting the decision-making process and lead to measures that provide greater equilibrium between the private and public interests. Another of the inconveniences is that the model requires precise input data to function, when the degree of accuracy of the available information is not normally known. This inconvenience is common to the other LUTI optimisation models and similar to other mathematical programming applications in general.

7.3.2 THE FIRST APPLICATION OF THE TECHNIQUE FOR OPTIMUM PLACEMENT OF ACTIVITIES INTO ZONES MODEL

The first application of the TOPAZ model was developed in 1970 by Brotchie and Sharpe in the city of Melbourne (Australia) in collaboration with the authority responsible for planning the development of the city (Brotchie and Sharpe 1975). The aim of the application was to find the best growth patterns over a future period of time. The initial study in Melbourne was performed on a macro-scale for which the Melbourne region was divided into 34 zones considering three types of land use:

- High-density housing
- Low-density housing
- Industrial and commercial use

Both public and private transport modes were considered, modelling flows of people to work and to other residential or industrial zones. The model was applied to minimise the total costs resulting from the combination of the urban development costs and transport costs during the periods: (1) 1970–1980 and (2) 1980–2000.

The objective function of the simplest version of TOPAZ (Brotchie et al. 1980) was used in the first Melbourne application to minimise the transport costs (first term) and the development activities cost (second term).

$$\min U\left(a_{ijm}\right) = \sum_{i}^{N}\sum_{j}^{M}\sum_{k}^{N}\sum_{l}^{M}\sum_{m}^{T}\sum_{n}^{Y} S_{ijklmn}R_{jlmn}B_{ijklmn} + \sum_{i}^{N}\sum_{j}^{M}\sum_{m}^{T} a_{ijm}c_{ijm} \quad (7.6)$$

s.t.

$$\sum_{i}^{N}\sum_{m}^{T} a_{ijm} \le z_j \quad \text{for } j = 1,2\ldots M \quad (7.7)$$

$$\sum_{j}^{M} a_{ijm} = A_{im} \quad \text{for } i = 1,2\ldots N; m = 1,2\ldots T; \; im = NT+1 \quad (7.8)$$

$$A_{NT+1} + \sum_{i}^{N}\sum_{m}^{T} A_{im} = \sum_{j}^{M} z_j \quad (7.9)$$

$$a_{ijm} \ge 0 \quad \text{for all } i, j, m \quad (7.10)$$

where:

$U(a_{ijm})$ is the total costs (costs less benefits) of distributing a_{ijm}

a_{ijm} is the part of an activity or total use of land A_{im} of type i distributed in zone j during time period m

S_{ijklmn} is the interaction between new activities i that are in zone j and new activities k that are in zone l, for transport mode n during time period m

R_{jlmn} is the journey time between zones j and l for mode n during time period m

B_{ijklmn} is the benefits less the costs of unit S_{ijklmn} along the length of a route between zones j and l

c_{ijm} is the benefits less the costs of establishing and managing an activity i in a zone j during the time period m

z_j is the area available for all the activities in zone j

N is the number of activities

M is the number of zones

T is the number of time periods

Y is the number of transport modes

The first constraint (Brotchie et al. 1980) guarantees that each zone receives one or various land uses, so no zone remains without being assigned some kind of land use. The second constraint guarantees that all the activities are distributed in the study area. The third constraint assures that the overall space being occupied by the different land use activities is equal to the capacity of all the zones, where A_{NT+1} is a dummy activity created to absorb the excess capacity between the zones and activities. Finally, the fourth constraint guarantees that the parts of land use or activities are positive.

All these restrictions are linear, whereas the objective function of the model is not linear, so the complete TOPAZ model is a non-linear programming problem that can be solved using different optimisation techniques.

Brotchie and Sharpe tested a series of planning alternatives changing variables in the model, the development of densities, the infrastructure costs, the public and private transport costs, traffic speeds, trip generation and vehicle occupancy. Of the results that were obtained from the TOPAZ application in Melbourne it is important to mention that the solution is optimum if the two time periods are considered simultaneously because the location of activities in the second period takes into account the journeys that occur with the location of activities in the first period.

The TOPAZ model was applied later in other Australian cities such as Gosford and Darwin as well as in cities in the United States, New Zealand, Iran and Indonesia.

In the United States, the model was applied for the development of planning in the New River Valley district in the south east of Virginia (Dickey and Najafi 1973). The district, including the cities of Floyd, Giles, Montgomery and Pulaski was divided into 40 zones and 5 types of land use activities: (1) residential, (2) commercial, (3) leisure, (4) public and (5) industrial. Furthermore, the zone is characterised by its peculiar orography: slopes, type of land, natural limits and artificial borders. These characteristics condition the location of the different kinds of land use. For example, the areas where the slopes are steeper than 20% or where flooding is common were not included in the study area. The model generates multiple solutions with maximum and minimum costs for journeys, construction and purchase of buildings, access to installations, supply of services and others. All these results were compared to generate solutions, which were low cost considering all the political and social factors involved.

Another application was performed in Tehran (Iran) in 1976 (Brotchie et al. 1980) to balance the maximisation of the overall benefit resulting from the resources used in introducing a transport system in the city with the maximisation of the benefits perceived by the public. An in-depth comparison was made between the infrastructure costs and the transport costs, resulting in a viable public transport plan, which also determined optimum land use patterns and predicted savings in the journey costs over future years.

Example 7.1: Application of the Technique for Optimum Placement of Activities into Zones Model

As an example of a simple application of the TOPAZ model, suppose a region is divided into three newly developed zones, an already developed zone and the urban centre (Figure 7.1). This example will consider only two types of land use: new housing development and new industrial development as well as only one mode of transport and one time period. Therefore, the objective function of the TOPAZ model is simplified to

$$\min U(a_{ij}) = \sum_i^N \sum_j^M \sum_k^N \sum_l^M S_{ijkl} R_{jl} B_{ijkl} + \sum_i^N \sum_j^M a_{ij} c_{ij} \tag{7.11}$$

The data required to solve it are shown in Tables 7.1 through 7.3. The first table shows the average journey times between the different zones with the assumption that the matrix is symmetrical. Table 7.2 shows the benefits less than the costs of establishing residential and industrial land use in the different zones, where the other zones are already completely developed. Table 7.3 shows the available development land in square meters for each type of land use in each zone.

The variables S_{ijkl} and B_{ijkl} are normally very complex functions, which require additional data and which may be dependent on the new land uses, which are developed in the different zones. This example has been simplified by assuming $S_{ijkl} = a_{ij} \cdot \alpha$ and $B_{ijkl} = a_{ij} \cdot \beta$, where α and β are variables representing all the additional data.

Furthermore, to apply the constraints to the objective function it has been considered that the available area for new land uses in zone 1 is 1.32 km², in zone 2 is 0.6 km² and in zone 3 is 0.35 km², and that the central and the already developed

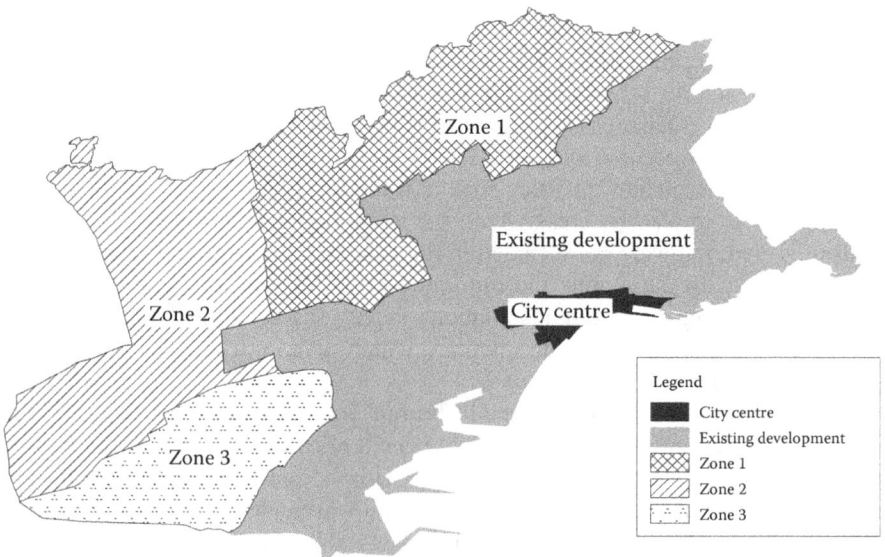

FIGURE 7.1 Solution of the TOPAZ model for the study area is divided into three development zones.

TABLE 7.1
Values of R_{jl} (min)

	CENTRE	DEVELOPED	Zone 1	Zone 2	Zone 3
Centre	3.2	12.8	11.2	18.4	18.8
Developed	12.8	13.2	12.4	16.4	18
Zone 1	11.2	12.4	8.8	14.8	18.4
Zone 2	18.4	16.4	14.8	6.4	9.2
Zone 3	18.8	18	18.4	9.2	4.8

TABLE 7.2
Values of c_{ij} (€/m^2)

	Zone 1	Zone 2	Zone 3
Housing	120	110	100
Industrial	90	80	70

TABLE 7.3
Maximum Values of a_{ij} (m^2)

	Zone 1	Zone 2	Zone 3
Housing	1,220,000	400,000	150,000
Industrial	100,000	200,000	200,000

zones do not have any available free space. The amount of residential and industrial land to be developed is 1.5 km² and 0.3 km², respectively.

The data described earlier are processed to find the land use shown in Table 7.4.

The data in Table 7.4 show that zone 3 is the optimum zone for industrial development as it already occupies all the area available for this land use, whereas none of the land in zone 1 is destined for industrial use due to its higher transport cost. Residential land use occupies all the allotted available land in zones 2 and 3; however, some land remains free for residential use in zone 1. The greater cost showed by zone 1 is mainly due to the higher costs associated with setting up either of the two activities compared with the other two zones.

TABLE 7.4
m^2 of New Land Use Developments

	Residential	Industrial
Zone 1	950,000	0
Zone 2	400,000	100,000
Zone 3	150,000	200,000

7.3.3 OTHER MULTI-OBJECTIVE OPTIMISATION MODELS

The Projective Optimization Land Use Information System (POLIS) is a version of the TOPAZ model that is aimed at helping planners in their evaluation of policies and predictions for future changes in population and employment (Prastacos 1986). In POLIS, the choice of housing depends on its availability and the journey to work. The location of the retail premises depends on the closeness to centres of population and attractive shopping areas. Finally, industry is distributed depending on the accessibility to work and the existence of a structure of production. It can be considered as a dynamic model as it simulates the change between the two situations. An increase in opportunity to employment and/or housing is assigned in each time period and the jobs are relocated with respect to the base year, increasing the number of jobs to be distributed. The location patterns of the model are similar to those of the base year because the constraints on the objective function make the model distribute all or part of the housing and jobs from the base year into specific zones.

Dökmeci et al. (1993) developed a multi-objective land use planning model based on the TOPAZ model. The model was applied to a hypothetical region to determine the optimum location of different land uses with two objectives that needed optimising. The first was to maximise income less costs involved with the urban land use, accessibility, installations and services, whereas the second objective minimises the sum of the weighted distance between the different land uses.

7.4 Bi-LEVEL OPTIMISATION MODELS

Bi-level optimisation models are used to optimise problems in which the two agents wish to maximise or minimise their objectives, one on the upper level and another on the lower level, where the agent on the lower level depends on the decision taken by the agent on the upper level, leading to an optimum decision, which benefits both the agents. Another of the characteristics of bi-level programming is that each of the levels, with their respective objective function and their pertinent constraints, optimises their net benefits independently as agreements between agents are not allowed. The problems of bi-level mathematical programming can be solved using meta-heurisitc methods such as genetic algorithms, ant colony algorithms, tabu search, the Stackelberg evolutionary algorithm and others.

Bi-level optimisation models have multiple applications (dell'Olio et al. 2006, Ibeas et al. 2009), including their application to LUTI models. Numerous researchers have applied bi-level optimisation models to address different goals linked to transport and land use.

The problem of land use and transport based on equity (ELUTP) (Lee et al. 2006) is solved using a bi-level optimisation problem to address whether the benefits of a network are shared equally between its users. The problem of land use development is considered in terms of changes in the equilibrium of journey costs between the origin–destination pairs. At the upper level, a multi-objective function is used to maximise the production of each residential zone under some physical constraints and equality constraints in which one user is not benefitted at the expense of another. The multi-objective function is transformed into one simple function by calculating the weighted sum of the journeys representing the costs and priorities in the development of land uses in the different zones. The lower level is characterised by the route choices, the origins and the destinations of the users of transport as a reply to the decisions of the transport planners.

Another example is the study by Chang and Mackett (2006), which tried to minimise the cost of transport and maximise the attractiveness of the location of different land uses. On the one hand, this model differentiates between the two types of population: (1) those who confront the decision to move residence and (2) those who do not carry out the change and their only interest is to minimise the transport cost. In this case, the upper level of the bi-level equilibrium model represents the behaviour of those who are willing to move from their home based on what they are willing to pay for the homes and on the probability of them being located between the origin–destination pairs. This probability is used to update the origin–destination matrix that is necessary on the lower level. On the other hand, the lower level considers the attractiveness of the different locations and the transport cost, which will be used on the upper level.

The study by Chang and Mackett (2006) is just an example; other researchers with similar goals have developed different optimisation models because the objective functions and the constraints of the model can change, as can the hypotheses or the variables being considered. The results of each one of the cases will, therefore, be different.

For example, Yim et al. (2011) specified a bi-level model that tried to distribute residential development and employment in an optimum way, so much so that they minimised the probability of overloading links on the network. The structure of their proposed bi-level model was as follows: the lower level solves the problem of the vehicle flow distribution and assignment, and the upper level maximises the trustworthiness index of the capacity of the network links with respect to the residential and employment assignments and network improvements. The hypothesis of this model states that the network planner has the necessary resources to increase the capacity of the existing roads on the network and the limitations in the material and financial resources as constraints.

Other models consider the interaction between the transport and the land use through time. The model proposed by Szeto et al. (2015) is a bi-level multi-objective optimisation model aimed at addressing the economic, social and environmental aspects involved in the design of a road network. In this case, the lower level solves the land use problem depending on time subjected to two constraints: (1) Lowry-type land use constraints and (2) constraints on the transport model through the choice of mode and route in time. The upper level considers sustainability indicators along with financial and design constraints. The sustainability indicators are such items as vehicle emissions, changes in the consumer surplus (the difference between what the users are willing to pay and what they actually pay for their journey), the variance in the discounted benefits of the real estate owners as a measure of their equity and the sum of the variation in the discounted user costs generalised for all the origin–destination pairs. The design and financial constraints considered are restrictions of capacity, tolls and cost recuperation with respect to incomes.

7.5 CONCLUSION

This chapter provides a historical overview about the different optimisation models applied in the field of LUTI modelling. The first LUTI optimisation model was developed by Herbert and Stevens and gave an optimal distribution of residential land uses in accordance with the theory of Alonso. However, the resulting distribution could be unrealistic because it did not consider the environmental constraints and diverse externalities present in urban areas. Nevertheless, this model has been the foundation for the development of all the later models, which have incorporated multi-objective optimisation techniques, such as the TOPAZ model and bi-level equilibrium models. The TOPAZ model provided greater flexibility and adaptability to variable situations, whereas bi-level equilibrium models are able to consider two agents that optimise different interdependent objectives. Optimisation models applied to LUTI modelling provide a support tool for the optimal location of activities by considering their interaction in space and are also useful for designing proposals and for comparing existing situations with an optimal solution.

REFERENCES

Alonso, W. 1964. *Location and Land Use: Toward a General Theory of Land Rent, Publications of the Joint Center for Urban Studies of the Massachusetts Institute of Technology and Harvard University*. Cambridge, MA: Harvard University Press.

Brotchie, J. F., Dickey, J. W. and Sharpe, R. 1980. *TOPAZ: General Planning Technique and its Applications at the Regional, Urban, and Facility Planning Levels, Lecture Notes in Economics and Mathematical Systems 180*. Berlin, Germany: Springer-Verlag.

Brotchie, J. F. and Sharpe, R. 1975. A general land use allocation model: Application to Australian cities. In *Urban Development Model*, Baxter, R., Echenique, M. and Owers, J. (Eds.), Lancaster, UK: Construction Press, pp. 217–236.

Chang, J. S. and Mackett, R. L. 2006. A bi-level model of the relationship between transport and residential location. *Transportation Research Part B: Methodological* 40 (2):123–146.

dell'Olio, L., Moura, J. and Ibeas, A. 2006. Bi-level mathematical programming model for locating bus stops and optimizing frequencies. *Transportation Research Record: Journal of the Transportation Research Board* 1971:23–31.

Dickey, J. W. and Najafi, F. T. 1973. Regional land use schemes generated by TOPAZ. *Regional Studies* 7 (4):373–386.

Dökmeci, V. F, Çagdas, G. and Tokcan, S. 1993. Multiobjective land-use planning model. *Journal of Urban Planning and Development* 119 (1):15–22.

Foot, D. H. S. 1981. *Operational Urban Models: An Introduction*. London, UK: Methuen.

Herbert, J. D. and Stevens, B. H. 1960. A model for the distribution of residential activity in urban areas. *Journal of Regional Science* 2 (2):21–36. doi:10.1111/j.1467-9787.1960.tb00838.x.

Ibeas, A., Moura, J. L. and Dell'Olio, D. 2009. Planning school transport: Design of routes with flexible school opening times. *Transportation Planning and Technology* 32 (6):527–544.

Lee, D.-H., Wu, L. and Meng, Q. 2006. Equity based land-use and transportation problem. *Journal of Advanced Transportation* 40 (1):75–93.

Markowitz, H. 1956. The optimization of a quadratic function subject to linear constraints. *Naval Research Logistics Quarterly* 3 (1–2):111–133.

Prastacos, P. 1986. An integrated land-use—Transportation model for the San Francisco region: 1. Design and mathematical structure. *Environment and Planning A* 18 (3):307–322.

Szeto, W. Y, Jiang, Y., Wang, D. Z. W. and Sumalee, A. 2015. A sustainable road network design problem with land use transportation interaction over time. *Networks and Spatial Economics* 15 (3):791–822.

Wegener, M. 1994. Operational urban models: State of the art. *Journal - American Planning Association* 60 (1):17–29.

Yim, K. K. W., Wong, S. C., Chen, A., Wong, C. K. and Lam, W. H. K. 2011. A reliability-based land use and transportation optimization model. *Transportation Research Part C: Emerging Technologies* 19 (2):351–362.

Section III

Steps for Building an Operational Land Use–Transport Interaction Model

This section explains the different steps that need to be taken to develop an operational land use–transport interaction (LUTI) model on an urban or regional scale. The following three points need to be considered before developing a LUTI model: (1) define the purpose of the model, (2) define the study area and (3) establish the data required to feed the model. An explanation of these points will be followed by a brief overview of the estimation of four of the main sub-models, which normally form the nucleus of land use and transport interaction models: (1) the population location model (Chapter 9), (2) the economic activities location model (Chapter 9), (3) the real estate pricing model (Chapter 10) and (4) the transport model (Chapter 11). All these sub-models will be illustrated through their practical application to the urban area of Santander, a medium-sized city located in the North of Spain.

8 Definition of the Model's Purpose, the Study Area and the Input Data

Rubén Cordera, Ángel Ibeas and Luigi dell'Olio

CONTENTS

This chapter introduces the steps that need to be taken before defining and estimating a land use and transport interaction model. These steps will try to answer the following questions: why do I need a land use–transport interaction (LUTI) model and what is it useful for? What area of study should I be modelling? How should I divide up my study area? What data do I need to correctly estimate my model? What confidence level can I place on the results that are produced by my model?

8.1 DEFINING THE MODEL'S PURPOSE

Perhaps the most important step to take before developing a LUTI model is to define if it is really necessary and what problems we are looking to help solve by applying it. As Lee (1973) stated, the use of LUTI models is not justified if simpler, cheaper and more effective tools are available to solve the same problems. Mathematical models are tools used by planners and analysts to try to address different questions more precisely about how a system behaves when it is changed in certain ways. They can also be used to evaluate strategic urban and transport plans in order to choose between different planning alternatives (Figure 8.1). However, they are complex tools with significant development and running costs that require the employment of specialised personnel meaning that their use must be clearly justified.

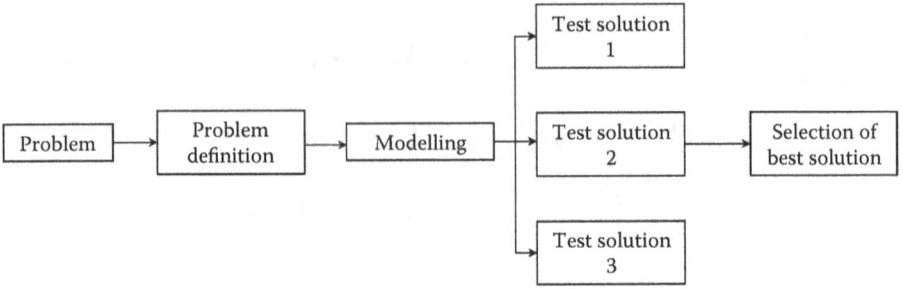

FIGURE 8.1   Use of modelling for evaluating solutions. (Based on Willumsen, L.G., *Better Traffic and Revenue Forecasting*, Maida Vale Press, London, UK, 2014.)

Given their nature, LUTI models can address different problems from those more suited to other types of models such as transport models, demographic models or purely economic models. Compared to the other models, LUTI models are characterised by their ability to spatially relate transport to land use that makes them particularly suitable to address questions such as the following (Foot 1981, Martínez 2000):

- Where will future urban development take place in a study area?
- What impact will new transport infrastructure (e.g. a motorway or a new type of railway) have on the future locations of population, activities and journey patterns?
- Up to what point is the location of urban agents influenced by the accessibility conditions of places?
- Does new transport infrastructure generate demographic and economic growth?
- What effects will new urban development have on transport?
- What will be the best location for a facility and how will it affect the rest of the urban area?
- How will employment decentralisation affect an urban area?
- Will the profits generated by increased accessibility to opportunities be capitalised on by real estate owners? By how much?

The choice of the main objective for the model should be the basic determining factor when choosing the type of model to use. This could range from an analysis of the impacts of different policies or support for different planning alternatives to specific questions that need answering. If this initial reflection is not made, then any later errors could be difficult to solve, or the incorrect model could be chosen, resulting in excessive financial or time costs for the desired objectives. Both issues can be avoided by establishing a list of problems or questions that need answering and by relating them to the kind of model that helps to address them.

Example 8.1: Defining the Purpose of a LUTI Model

A real case study centred on the urban area of Santander will be used to highlight the points explained in this section. Santander is a medium-sized city, capital of the region of Cantabria in Northern Spain. The city is located about 100 km west of metropolitan Bilbao and 410 km north of the capital of Spain, Madrid. The city of Santander currently has a population of 180,000 inhabitants, and around 260,000 people live within its sphere of influence.

A LUTI model was proposed as the best option to simulate how changes made to the transport system would affect the rest of the urban system. The main goal was to evaluate the impact of introducing new modes of public transport and policies to address the restriction of private car use in the city centre. Some of the actions to be addressed were the introduction of a light railway service, the introduction of a bus with a high level of service in the central urban corridor, the restructuring and improvement of the existing public transport service, the repercussions of congestion charging to enter the city using a private vehicle and the repercussions of expanding a charging system for on-street parking. The goal was not only to simulate the impacts of these measures on the transport system but also on the phenomena related to land use such as urban accessibility, the propensity for urban development to take place in certain areas and the changes in real estate prices. A LUTI model was seen as the best alternative rather than other more specific and restricted models, given the strong interrelationship between transport and land use in all the stated cases. It was decided that the LUTI model had to be built around an already existing transport model for the city. The land use models had to provide results related to changes in system accessibility caused by the introduction of stated policies, had to calculate the changes in the locations of population and activities and had to estimate how these changes would impact the affected real estate. There was no desire to use an overcomplicated model with a high demand of data requirements, as data availability was limited in the study area and as no previous attempt had been made to apply this type of model.

8.2 DEFINING THE STUDY AREA AND ITS ZONING

The study areas that are normally simulated by LUTI models are differently scaled urban zones ranging from small- and medium-sized cities to large regional urban areas. They are, therefore, generally quite ample areas made up of a system of locations and displacements in relatively closed environments. The main aim in defining the study area is that the interactions, the journey and location patterns are much more probable internally than to external zones. Where numerous interactions occur with external areas, LUTI simulation would not be very accurate because it would depend on journeys and patterns that are not endogenously explained by the model. Therefore, in urban nuclei, it is better to not only model the most consolidated urban space, the space normally contained within an administrative area, but also all the areas within its sphere of influence where commuting is a normal occurrence. In a highly populated urban area, the preferable modelling zone would be the complete metropolitan space. However, this external limit around the study area will normally be conditioned by the existing administrative divisions as they will normally provide the available information from statistical sources such as the population census.

The function behind zoning the study area is to aggregate households, economic activities and the built space corresponding to each area (housing, retail premises, facilities and others) in usable sized portions of territory. It also provides a division that facilitates data interpretation and that quickens the solution algorithms associated with the different models. A system of zoning created for a land use model should be designed considering the following criteria (Foot 1981, de Dios Ortúzar and Willumsen 2011):

- The zoning system should be compatible with the existing administrative divisions, particularly with that used by the current census. This criterion is probably the most important as a lot of demographic and economic data are available only on a census level.
- The zones need to be as homogenous as possible in their composition of activities and population. Clearly differentiated zones must never be aggregated with others however small they are.
- It is important that zonal limits are compatible with pre-existing zoning.
- Population differences between zones that have occurred due to recent development need to be considered.
- Zones should not be too large as this means that a lot of the interaction takes place within each zone rather than between zones.
- However, the zones should not be too small to avoid the location and interaction patterns becoming too disaggregated and difficult to describe. Too many zones also make it difficult to interpret the results provided by the models.

Where different systems of land use zoning and transport zoning are available, they should both be completely compatible. It is a common practice to use a more detailed zoning for modelling transport and a more aggregated zoning for simulating the location of employment and population as mobility patterns can normally be modelled with greater accuracy.

Some models have also opted to use cells as a unit of observation and even urban plots because they have the specific data available for the study area where the model is being applied (Waddell 2002). However, a division made on a zonal level continues to be the most commonly used, and frequently it is the only one that is feasibly applicable in many study areas, given that the data are only available on a census level.

Example 8.2: Zoning and Definition of the Study Area for the Application of a LUTI Model in Santander

Figure 8.2 shows the study area that was finally defined for the application of the LUTI model to Santander. The city of Santander is located in the north of the area. The delimited area is divided into a total of nine municipal administrative units.

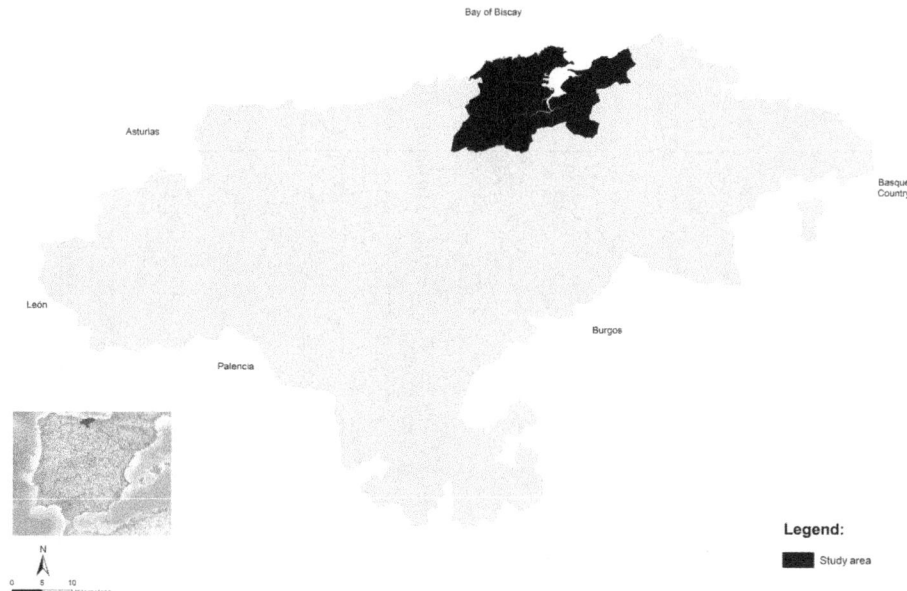

FIGURE 8.2 Location of the study area.

However, around the urban nucleus of the city of Santander, multiple relationships occur with neighbouring areas and suburban developments, resulting in an important number of commuting journeys. These phenomena meant that the study area is needed to be extended beyond the city itself towards neighbouring settlements around the Bay of Santander, where second homes and leisure activities are also located. Other urban areas in the region, particularly the town of Torrelavega (pop. 50,000) about 30 km from Santander, are considered to be part of the external area. Another possibility would have been to include Torrelavega inside the study area, but that would have come with added complications; in that, the town has its own sphere of influence that would also have to be added to the model. Therefore, it was finally decided to adopt an intermediate solution that did not excessively extend the study area to avoid the need of collecting more data that would inevitably prolong the model's calibration time.

Figure 8.3 shows the final zoning used for the study area. The urban nucleus of the city can easily be identified, given the greater zonal disaggregation of that area. Normally, a more disaggregated zoning is used in those areas, where there is greater diversity of population and activities and where more detailed results are required depending on the goals of the model.

A total of 42 zones were created starting from the most disaggregated information possible. In this case, the information was obtained from the population and housing census that, in the case of Spain, does not go over 2000 inhabitants per section. The sections in Santander were smaller because of the higher population density. All the delimited zones were designed to have a certain internal homogeneity and an intermediate size that were sufficient for the conditions and goals of the modelling process.

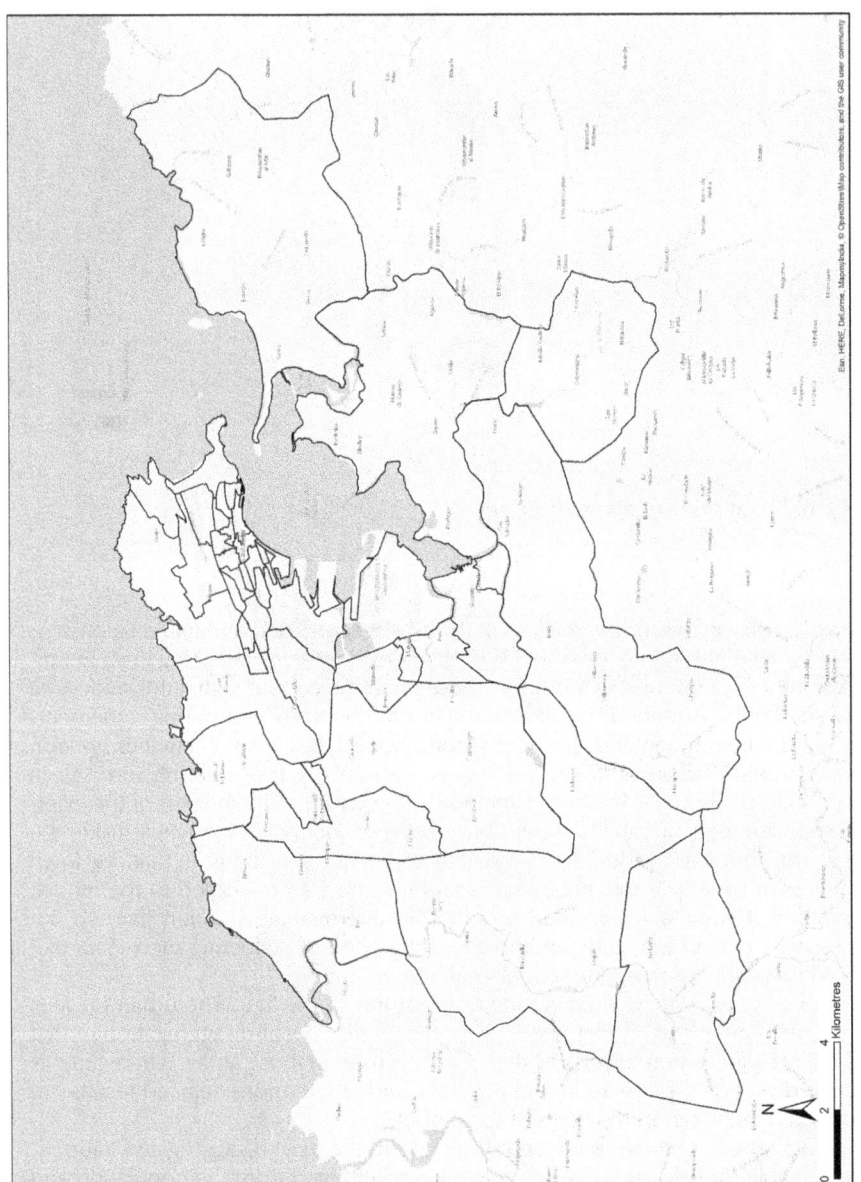

FIGURE 8.3　Zoning of the Santander urban area.

8.3 DEFINING THE DATA REQUIRED FOR ESTIMATING THE MODEL

Once the reason behind the model, the study area and the chosen zoning has been defined, the next logical step is to start collecting the data required to feed the model. The data will depend on the size of zoning and on the type of model chosen; the availability of the data will also condition the size of the study area and the chosen zoning. It will always be more convenient to collect the data on the most disaggregated and detailed zonal level possible and to proceed in aggregating them where necessary.

Normally, in all countries one of the basic data sources will be the population and housing census. According to the United Nations definition (United Nations 2008b), a population and housing census is the process of collection, compilation, evaluation, analysis and publication of demographic, economic and social information in a specified period of all people and housing in a delimited territory. So, a census does not only collect demographic and housing information but, depending on the country, they can also contain important social and economic data and even information about the mobility patterns of the population. This makes the census a basic source of information that can, however, be somewhat out of date due to the timescales of data collection. The United Nations recommends holding a census at least every 10 years, meaning that during the period between collection and publication of the new census, there could be an important difference between reality and available data. On many occasions, it will be necessary to complement the census information with intermediate estimations that update the information where available. If this does not happen, then other sources of information will need to be used such as population counts carried out on a municipal or local administrative scale. Normally, both sources provide data with an ample degree of disaggregation per territorial unit in thousands or hundreds of inhabitants.

Depending on the characteristics of the census, certain economic information such as the location of employment may not be available. Where this is the case, it will be necessary to recur to alternative administrative sources, such as company census or specifically designed surveys to collect that type of information. It may sometimes be useful to combine the collection of specific data using samples with other planning or research activities such as origin-destination surveys for mobility planning.

The need to collect data from different sources can generate problems that are derived from different spatial and temporal frameworks adopted by different sources. Furthermore, the measuring tools used in each of the statistical operations do not necessarily need to be homogenous, given that different phenomena can be defined in different ways. However, as these problems are not easily solvable, the modeller needs to opt for a compromise solution and tries to collect all the data with the fewest possible discrepancies and required number of fits. If the data are seen to be very deficient, the modelling team should question whether they should be working at this level of detail or whether they can work with a LUTI model that requires less complicated data. Another more expensive possibility available to the modellers is to collect all the information directly via sampling.

Some of the kinds of data normally required by LUTI models can be seen in Figure 8.4. This data can be divided into four basic categories:

- *Sociodemographic data about the population*: As previously mentioned, this is normally the most easily accessible and available data provided by the population and housing census. The model may require the data to be disaggregated according to characteristics such as age groups or income levels. Another source, which could be of interest, is household surveys that address family budgets because they can provide information about places and shopping-related trips.
- *Data about economic activities*: These data can sometimes be difficult to obtain as alternative sources to the census need to be used, for example, company directories. Activity location models need to have employment location data available, and this data can often be imprecise, especially given that a company can have more than one physical facility in which a variable number of jobs are located. Furthermore, activities should be classified by economic sector if the aim is to disaggregate the modelling for different kinds of employment. Classifications according to activities are complex because the same company can often be categorised in multiple economic sectors. The most common classifications on a statistical level are the Statistical Classification of Economic Activities in the European Community (NACE), the International Standard Industrial Classification (ISIC) of the United Nations (United Nations 2008a) and the North American Industry Classification System (NAICS) for the United States.
- *Real estate data*: If the impacts of transport on real estate prices need to be modelled, a basic requirement will be the availability of a real estate database about property characteristics and market prices. This type of data may not be available in many study areas, so specific surveys or data collection from online sources such as real estate websites may be required, even though the information obtained in the latter case will almost certainly be of lower quality.
- *Population mobility and activities data*: In many cases, the population census provides information about the most frequent trips made by household members for purposes such as work or study. Nevertheless, these data are normally insufficient for modelling the mobility of the population, so alternative sources will be required such as origin–destination and mobility surveys usually organised by the local transport authorities.

It may also be of interest to consider information on land use, built surface and retail surfaces in the different zones in the study area. This type of data can be found in cartographic sources in geographic information system (GIS) format or from the various and increasingly available servers, providing geographical data on the Internet.

All this data will then be used in a later phase to calibrate the parameters of the model's equations to provide as good fit as possible to the real observations. The technique for calibrating the parameters largely depends on their number and the

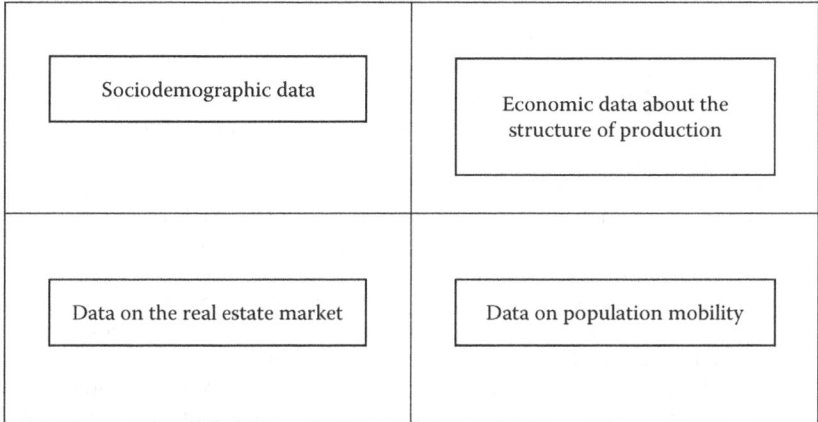

FIGURE 8.4   Data usually used by LUTI models.

functional form that are chosen. Theoretically, it could be done through a trial and error procedure in which some parameters are inserted into the equation, and the result is checked repeatedly until finding the values that provide the closest solution to the observed data. However, given that this process would be very slow, these parameters can be estimated using statistical techniques such as the minimisation of squared errors in linear models or the maximisation of likelihood, or the probability of observed data. These techniques also count with additional indicators and tests that can compare different models with each other, thereby facilitating the fit and the choice of the best model.

Apart from calibrating or estimating the parameters, the models also have to be validated. This process consists of checking the validity of the model by running it using different data from those that are used in the calibration/estimation process. Another alternative is to check the model against a passed and known state of the system, a process that is denominated as retrodiction.

Example 8.3: Data Required for Calibrating the LUTI Model of Santander

The input data can be deduced from the general requirements of the LUTI model used in the Santander case study. This data should be collected from the available resources in the study area. Specific surveys can also be designed and administered to collect it. Table 8.1 summarises some of the data that could be required to feed a LUTI model formed by a residential location submodel, an economic activity location submodel, a real estate pricing submodel and a transport submodel.

Residential location model: Data on the effective location of the population within the chosen zoning system are needed to feed a residential location model. If more precision is required from the model, then household or even individual data are needed on the socioeconomic characteristics of each of the observations for their segmentation into different categories.

The availability of accurate housing data is also important, how many homes are being supplied or how much residential built space there is, either to act as a location constraint in each area or to estimate a housing supply model. The most usual sources for obtaining this type of information would be the population and housing census and, where available, land use maps.

Economic activities location model: A basic piece of data required by these models is to know the number of jobs present in each of the studied zones. Typical sources for this type of information are the population and housing census and more specific administrative sources such as company directories, chamber of commerce and so on. As already highlighted, it is better if the companies are characterised following some standard classification to make them more easily segmentable.

Real estate pricing model: This type of model depends on the availability of a representative housing sample together with their market prices, their location and their structural characteristics (rooms, surface area, availability of garages etc.). If purchase prices are not available, then offer prices can be used as housing is normally overvalued by a certain variable percentage, depending on the study area. This kind of information can also be found on specific databases that deal with housing market transactions.

Transport model: Transport models are composed of a demand submodel, which simulates the trip generation, distribution and modal choice of users in each of their journeys, and a network submodel, which captures the structure of the links and nodes on the road network. This second model may be digitalised from cartographic sources, whereas

TABLE 8.1
Data Required for the Calibration of a LUTI Model

Model	Exogenous Input Variables	Available Sources
Residential location model	Residents segmented according to socioeconomic characteristics per zone	Population and housing census
	Surface area of housing per zone	Population and housing census/ land use mapping
	Number of houses per zone	Population and housing census
Economic activities location model	Employment by economic sector and zone	Population and housing census/ directory of companies and establishments
Real estate pricing model	Prices and characteristics of real estate by zone	Housing statistics
Transport model	Household data on trip generation, distribution and modal choice	Mobility survey
	Transport network	Mapping
	Information about transport network links	Local government information

the demand model requires specific data about journey patterns from a representative sample. For that, it is normally necessary to ask a specific mobility survey to collect the typical trips made by a population, considering origins, destinations, transport modes used and so on.

For illustration purposes and to present the notation that will be used in other chapters, a simplified case will be presented only for the city of Santander that is divided into eight large zones (Figure 8.5).

Table 8.2 summarises the population, jobs and housing present in each of the eight zones.

The totals for population, housing and jobs in the area are, therefore as follows:

$$P = P_1 + P_2 + P_3 + P_4 + P_5 + P_6 + P_7 + P_8 = \sum_i P_i = 177010 \qquad (8.1)$$

$$E = E_1 + E_2 + E_3 + E_4 + E_5 + E_6 + E_7 + E_8 = \sum_i E_i = 67659 \qquad (8.2)$$

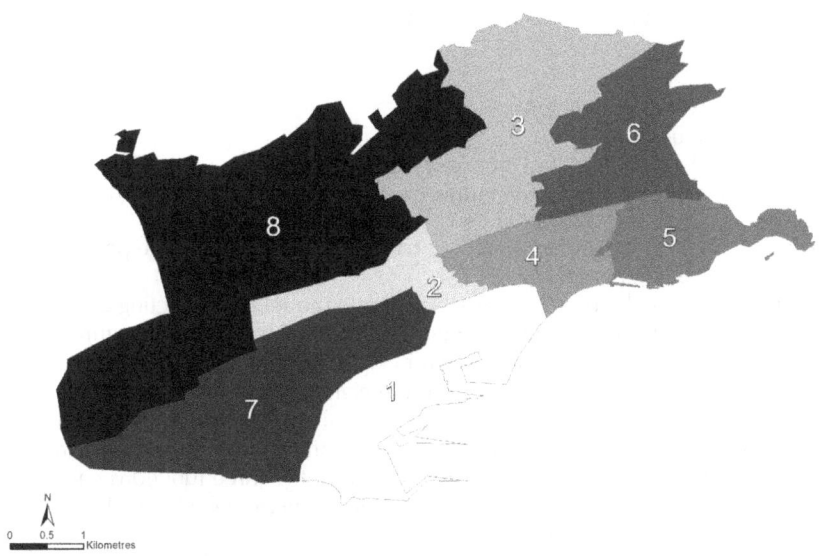

FIGURE 8.5   The city of Santander divided into eight zones.

TABLE 8.2
Characteristics and Simplified Zoning of the City of Santander

Model	Zone 1	Zone 2	Zone 3	Zone 4	Zone 5	Zone 6	Zone 7	Zone 8
Population	34,975	29,100	4,990	49,560	18,860	10,420	21,845	7,260
Jobs	17,706	5,304	952	21,847	4,948	4,932	9,286	2,684
Housing	18,185	13,970	2,495	26,075	11,205	6,195	10,880	3,450

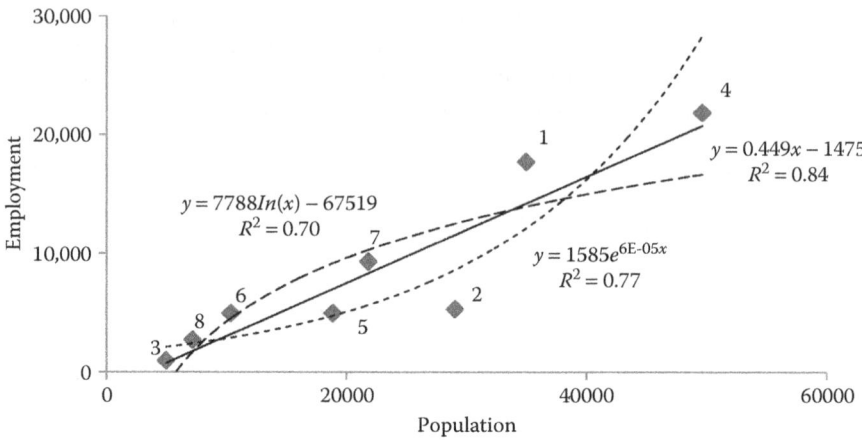

FIGURE 8.6   Simplified models for predicting employment as a function of population.

$$V = V_1 + V_2 + V_3 + V_4 + V_5 + V_6 + V_7 + V_8 = \sum_i V_i = 92455 \qquad (8.3)$$

where:

   P is the overall population

   V is the housing total

   E is the jobs total P_i, V_i and E_i correspond to the population, housing and jobs in specific zone i, which in this case can take the values of 1–8.

So in the area of Santander, there is a ratio of $P/V = 1.915$ people per home and $P/E = 2.616$ people per job.

   These data can be used to estimate a simple model for predicting employment in a zone based on some knowledge being available about the future population. Three functional forms are shown in Figure 8.6 that are fitted to the population and employment data observed in the eight city zones: a linear function, an exponential function and a logarithmic function. The estimation process provides the parameters corresponding to the intercept and to the population variable in each one of the functions. The goodness of fit of the three functions can also be checked using the R^2 indicator. For this example, the model with the best fit is the linear function with an R^2 of 0.84.

8.4   USE OF LUTI MODELS FOR MAKING PROJECTIONS AND PREDICTIONS

The results provided by LUTI models can be used for making projections and predictions about the future state of the urban area being studied. A projection is an extrapolation of the current situation based on a conditional argument about how certain key aspects of the urban system are going to behave. For example, how will urban development evolve if the population has a stable annual growth rate of 2%? Whereas a prediction is a statement about the most probable future state, in other words, it is based on the selection of the conditional factors that the analyst believes most probably will affect the development of the system. So, for example, the modeller can propose that future development of an urban area will occur in certain parts and not in others, given a probable annual population increase of 1%. It is normally assumed in both types of simulations that the calibrated parameters in the models will remain stable in the future. These parameters correspond to phenomena such as the preferences of the agents for different transport choice variables and locations. The prolonged stability of these parameters is an additional hypothesis of the model that can only be expected to be partially true.

The ability of a model to provide precise and accurate projections and predictions must also be established (Willumsen 2014). Precision is understood to be the level of detail that a model is able to provide, for example, in terms of spatial or temporal resolution, whereas the accuracy of a model refers to the difference between the calculations of a phenomena's characteristic and its real value. So a precise model will provide a very detailed result, but it does not need to bear any resemblance to reality, whereas an accurate model can provide a realistic, although less detailed, result. Therefore, an aggregate model can provide a projection or prediction with a reasonable fit to the real result but at a lower level of detail than the modeller requires on a zonal level or when using division according to activities. Precision and accuracy should be balanced although it is generally preferable to use a model that provides less precise and more accurate results.

Two large groups of variables can be differentiated in all models: exogenous and endogenous variables. Exogenous variables are the input data for the model that need to be updated externally to make projections and predictions using the model. Endogenous variables, on the other hand, are dependent model variables and are therefore calculated as part of the results of the simulations themselves (e.g. the number of people located in each zone of the study area). These results are the product of two overall types of modelling that allow individual models to be classified into three large groups: equilibrium models, quasi-dynamic models and completely dynamic models. The equilibrium models assume an initial state for the system and later introduce an external factor that takes the system onto a new point of equilibrium as can be seen in Figure 8.7a. Therefore, they are comparative equilibrium models as they do not really model the process by which the system adapts from the initial equilibrium to the final equilibrium. That final state of equilibrium should, therefore, refer to a date that is dependent on the values introduced into the exogenous variables. On the other hand, the dynamic models consider the presence of inertia in the system that makes the effects of any change to occur with a variable time delay, where it is also possible to differentiate between short-, medium- and long-term effects (Figure 8.7b). These kinds of models not only simulate the final equilibrium solution but also the process involved in arriving, thereby considering common frictions and resistances within urban change. Furthermore, the different rates of change occurring in the simulated phenomena must also be considered as, for example, aspects of the transport system like trip distribution and modal choice are more flexible and change more quickly than others related to land use, such as residential choice (Simmonds et al. 2013). In fact, the same factors may have differential effects on different characteristics of the system, and consecutive factors of change could change or neutralise the effects of previous factors (Figure 8.7c). In the quasi-dynamic kind of model, the results of an initial period of simulation are successively introduced into a following step until the final

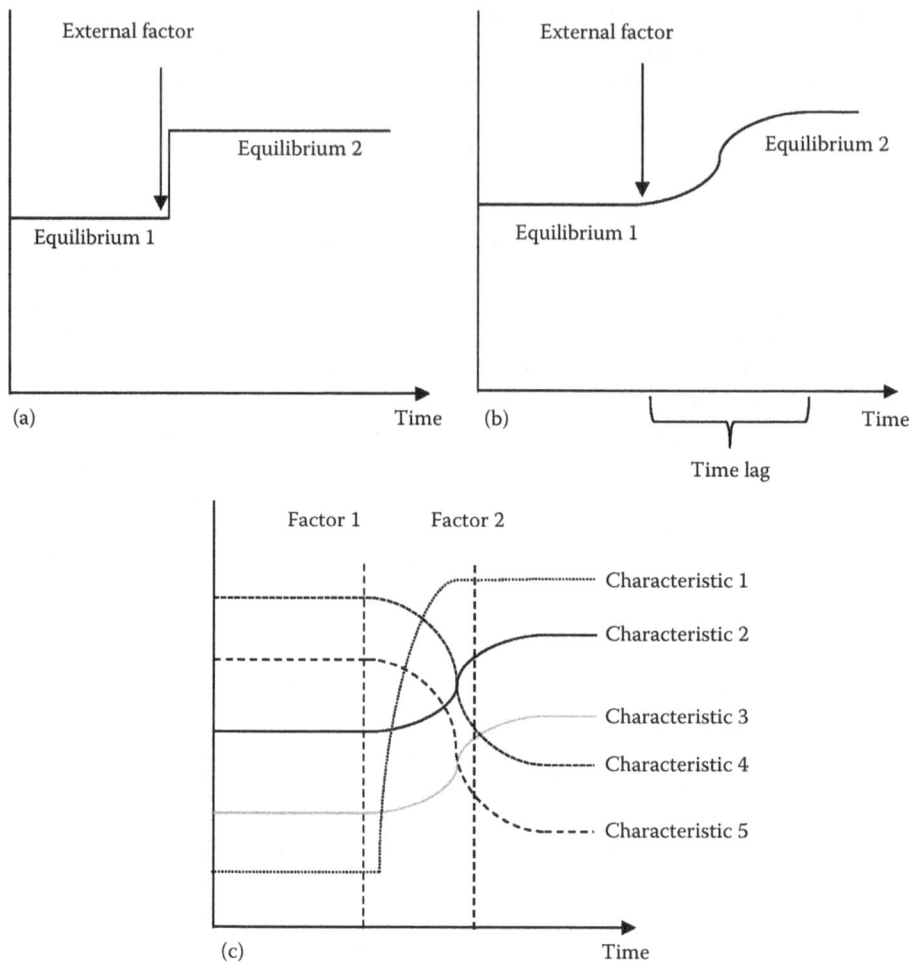

FIGURE 8.7 Type of equilibrium in a system being studied. (a) Equilibrium model, (b) quasidynamic model and (c) completely dynamic model.

solution is reached. Each step of the simulation is, therefore, used as the input for the next step, giving the model a certain dynamic component and allowing the transformation from initial to final equilibrium to be examined.

After obtaining the results from the model, it must be realised that these results must always be interpreted with caution. LUTI models are the tools available for the evaluation of urban policies and planning, but under no circumstances should they be seen as providing the definitive final answer to the questions being asked. The complexity of the models themselves, the simplified hypothesis adopted at a theoretical level, the difficulty of measuring some variables, the uncertainty about future processes that may or may not affect the system and many other factors mean that the results from the model should be interpreted more like trends within the system than as precise and accurate results. Although models will only offer partial answers to the questions being asked, the answers they provide could be of great help for professionals and decision-makers in structuring their ideas about the most plausible future scenarios and in helping them to detect the most probable patterns of change in the system being studied.

REFERENCES

de Dios Ortúzar, J., and L. G. Willumsen. 2011. *Modelling Transport.* Chichester, UK: John Wiley & Sons.

Foot, D. H. S. 1981. *Operational Urban Models: An Introduction.* London, UK: Methuen.

Lee, D. B. 1973. Requiem for large-scale models. *Journal of the American Planning Association* 39 (3): 163–178.

Martínez, F. J. 2000. Towards a land-use and transport interaction framework. In *Handbook of Transport Modelling,* edited by D. A. Hensher and K. J. Button, pp. 145–164. Oxford, UK: Elsevier Science.

Simmonds, D., P. Waddell, and M. Wegener. 2013. Equilibrium versus dynamics in urban modelling. *Environment and Planning B: Planning and Design* 40 (6): 1051–1070. doi:10.1068/b38208.

United Nations. 2008a. *International Standard Industrial Classification of All Economic Activities.* Edited by Economic and Social Affairs. Statistics Division. New York: United Nations Publication.

United Nations. 2008b. *Principles and Recommendations for Population and Housing Censuses.* Edited by Department of Economic and Social Affairs. Statistics Division. New York: United Nations.

Waddell, P. 2002. UrbanSim: Modeling urban development for land use, transportation, and environmental planning. *Journal of the American Planning Association* 68 (3): 297–343.

Willumsen, L. G. 2014. *Better Traffic and Revenue Forecasting.* London, UK: Maida Vale Press.

9 Models for Simulating the Location of Population and Activities

Eneko Echaniz, Rubén Cordera and Luigi dell'Olio

CONTENTS

9.1 INTRODUCTION

Population and activity location models are aimed at simulating the spatial distribution of households and companies, normally within an urban area. This kind of model is able to estimate the number of resident households and the number of jobs available in each of the zones into which a study area has been divided.

This chapter will provide a review of the state of the art in the field of modelling the location of urban agents within a territorial system, starting with an overview of the main residential location models found in the literature. A practical example applied to the city of Santander (Spain) will also be presented to illustrate the estimation and how this type of models works. This will be followed in the second section by various economic activity location models, which will also include practical examples. In both cases, residential location and economic activity location, the model being applied is based on random utility theory (see also Chapter 6).

These days, integrated models are available that are capable of simultaneously calculating the locations of population and economic activities. However, because the evolution of the two types of models has not been altogether parallel, the differentiation between the location of residential and economic activities continues to be thought necessary and will be maintained throughout this chapter.

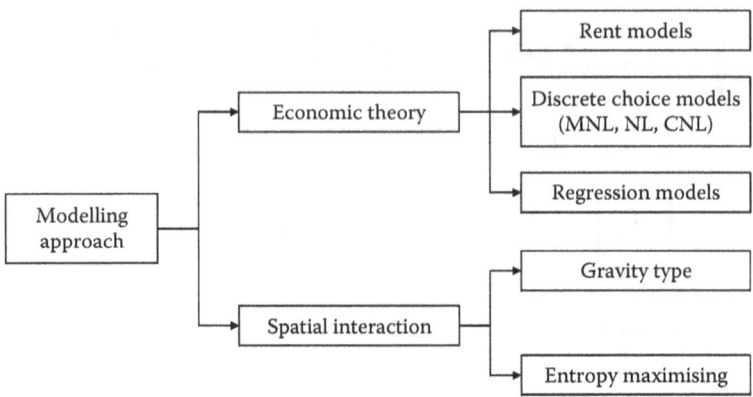

FIGURE 9.1 Classification of spatial location models. (Based on Pagliara, F. et al., *Residential Location Choice: Models and Applications*, Springer Science & Business Media, Berlin, Germany, 2010.)

Location models were mainly developed from two theories that appeared during the nineteenth century (Figure 9.1). The origins of the first of them, based around micro-economics, can be found in the work of von Thünen (1826) on the use of agricultural land (see Chapter 4). The second, based on the idea of spatial interaction, which, due to its analogy with the theory of gravity, led to the development of the so-called gravity models, was initially applied to migrant and commercial movements (see Chapter 5) in works such as Ravenstein (1885), Reilly (1931), Zipf (1949), Lill (1891), Young (1924), Bossard (1932) and Ikle (1954). However, none of these theories was initially applied to the phenomenon of residential location. Actual residential location models as we know them today had their origins in the work of Alonso (1964), who brought von Thünen's economic theory into the realms of urban residential location. Hansen (1959) and Lowry (1964) applied the principles of spatial interaction to residential location and, in the case of Lowry, also to the location of economic activities.

Later on, Wilson (1970) generalised the idea of spatial interaction based on the concept of entropy maximization (Senior and Wilson 1974). This new theory led to the development of more complex models in which the interactions could be structured by restricting the maximisation process.

Under certain conditions, econometric models based on discrete choice have been shown to match the maximum entropy models (Anas 1983). However, given the more disaggregated nature of the choice models and their greater ease of specification and interpretation, they have grown in stature in the field of operational land use–transport interaction (LUTI) modelling.

A summary of a variety of residential choice models applied in the evaluation of transport and land use policies and projects is presented in Section 9.2. A specific application will also be presented in order to illustrate the process of estimating these models. A summary of various activity location models is presented in Section 9.3 to provide a representation of the state of the art, and once again there will be an illustrative practical application for the case of Santander.

9.2 RESIDENTIAL LOCATION MODELS

A great variety of residential location models, developed according to the needs of the various research groups involved, can be found in the relevant literature. A short series of representative models that have been widely applied in different fields of research are presented in Sections 9.2.1 through 9.2.4. They were also chosen for the methodology used in their development and their possible future application.

9.2.1 DRAM MODEL

The DRAM location model was created by Putman (1979) and began its development in the 1970s, even though it continues to be used in the present day in multiple urban areas. DRAM and its successive evolutions (METROPILUS) are considered to be the residential choice model that is most widely applied in the United States (Oryani and Harris 1996). The DRAM model belongs to the ITLUP software package for analysing the interaction between transport and spatial household and company location patterns. The residential location model is complemented by the EMPAL employment location model.

The DRAM model is based on spatial interaction theory. At the time when the model was being developed, this theory was starting to be investigated as a consistent mathematical structure based on the theory of entropy maximisation.

The model is characterised by the inclusion of a multi-variate function to define a zone's attraction and a multi-parametric journey cost function. The attraction of zone A_j is estimated by using a Cobb–Douglas function that considers all the available attributes of zone j X_{nj} to the power of each one of the parameters to be calibrated (α,β etc.):

$$A_j = X_{1j}^{\alpha} X_{2j}^{\beta} X_{3j}^{\gamma} X_{4j}^{\delta} \tag{9.1}$$

DRAM is an interaction model, singly constrained to destinations, which is very similar to a multinomial logit (MNL) discrete choice model. The model also possesses a balanced constraining feature that can establish specific limits to the location of users in the different zones. DRAM can be used to locate up to eight categories of households defined by their income levels with individually estimated parameters. The structure of the model is presented as follows (Pagliara et al. 2010):

$$N_i^n = \eta^n \sum_j Q_j^n B_j^n W_i^n c_{i,j}^{\alpha^n} \exp(\beta^n c_{i,j}) + \left(1.0 - \eta^n\right) N_{i,t-1}^T \tag{9.2}$$

where:

$$Q_j^n = \sum_k a_{k,n} E_j^k \tag{9.3}$$

and

$$B_j^n = \left[\sum_i W_i^n c_{i,j}^{\alpha^n} \exp(\beta^n c_{i,j})\right]^{-1} \tag{9.4}$$

and

$$W_i^n = \left(L_i^v\right)^{q^n} \left(x_i\right)^{r^n} \left(L_i^r\right)^{s^n} \prod_{n'} \left(1 + \frac{N_i^{n'}}{\sum_n N_i^n}\right)^{b_{n'}^n} \tag{9.5}$$

where:

N_i^n is the type n households residing in i

$N_{i,t-1}^T$ is the type n households residing in i at iteration $t - 1$

E_j^k is the type k employment in zone j

L_i^v is the developable urban land in zone i

L_i^r is the residential land in zone i

$x_i = 1$ plus the percentage of developable land already developed in zone i

$a_{k,n}$ is the regional coefficient of the number of type n households per each type k worker

$c_{i,j}$ is the journey time or cost between zones i and j

$\eta^n, \alpha^n, \beta^n, q^n, r^n, s^n, b_{n'}^n$ are the empirically estimated parameters

The residential location model (Equation 9.2) is seen to be the sum of two terms. The first term corresponds to the sum of all the kinds of employment present in zones j (Equation 9.3), the costs of residing in zone i to make the journeys to work in zone j (Equation 9.4) and the attractiveness of each zone i to attract residents (Equation 9.5). The second term corresponds to the number of households that had already located in the zone during the previous step.

9.2.2 DELTA MODEL

The DELTA model began to be developed during the 1990s by the David Simmonds Consultancy. The model was created with the aim of making predictions in urban and regional zones, in particular to analyse the changes that occur as a result of the impact of different measures taken in the field of transport. The foundations of the first DELTA model can be found in Simmonds (1999) with more recent and complementary information available in Simmonds and Skinner (2003) and Simmonds and Feldman (2005).

The DELTA model presents a setup based on temporal 1-year cycles in which residential location is analysed considering the available space in each zone. At the same time, it develops a regional scale economic model that serves to discover the current levels of migration in the area, as can be seen in Figure 9.2.

The DELTA residential location sub-model is simultaneously a location–relocation model and a real estate market model. The changes occurring in location respond to five variables: accessibility, local environmental quality, number of available houses, quality of housing and the housing consumption utility. This latter variable is related to the proportion of salary that users are prepared to spend on services relating to their home.

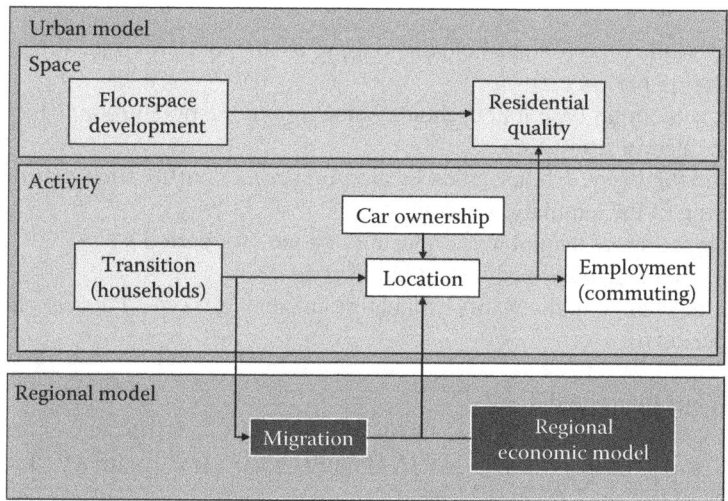

FIGURE 9.2 DELTA model using 1-year cycle. (Based on Pagliara, F. et al., *Residential Location Choice: Models and Applications*, Springer Science & Business Media, Berlin, Germany, 2010.)

To establish user housing choice, the model requires a detailed classification of each household: household composition, age of residents, employment situation of residents and their socioeconomic level. It is assumed that a change in residence is related to a change in some of the representative variables of the household, for example, a change in the composition of the family, a change in employment status or other factors. Once a household is considered as a candidate for changing location, it enters into the relocation process.

The location process divides households into two groups: new households, in other words, those that do not have previous residency in the study zone, and the mobile households that were already established as residents in the zone. The number of households required to locate using a transition model is initially calculated. This is followed by applying the inter-zonal migration model that establishes the number of households who move from one of the system's zones to another. The households that change zone are removed from the location groups of their initial zone and are then added to the pool of the destination zone. The aim of the location model is to establish residency for all the households present in both groups.

The equations governing the location model are based on weighted incremental logit models, with slight variations between the equations used for the pool group or the mobile group. For the group of new households:

$$H(LP)^h_{pi} = H(P)^h_{pa} \cdot \frac{H^h_{ti}\left(F(V)^H_{pi} / F^H_{ti}\right) \cdot \exp(\Delta V^h_{pi})}{\sum_{i \in a} H^h_{ti}\left(F(V)^H_{pi} / F^H_{ti}\right) \cdot \exp(\Delta V^h_{pi})} \tag{9.6}$$

where:

$H(LP)^h_{pi}$ is the type h households belonging to the pool group locating in zone i during period p

$H(P)^h_{pa}$ is the total type h households belonging to the pool group, which will be located in study area a

H^h_{ti} is the total type h households located into zone i during time t (at the beginning of the simulation period)

$F(V)^H_{pi}$ is the space available for housing in zone i for period p

F^H_{ti} is the housing previously occupied in zone i

ΔV^h_{pi} is the change in the utility of locating in zone i for type h households during period p

For the mobile households:

$$H(LM)^h_{pi} = \left\{ \sum_i H(M)^h_{pi} \right\} \cdot \frac{H(M)^h_{ti}\left(F(V)^H_{pi}/F(M)^H_{ti}\right)\cdot\exp\left(\Delta V^h_{pi}\right)}{\displaystyle\sum_{i\in a} H(M)^h_{ti}\left(F(V)^H_{pi}/F(M)^H_{ti}\right)\cdot\exp\left(\Delta V^h_{pi}\right)} \quad (9.7)$$

where:

$H(LM)^h_{pi}$ is the type h mobile households located in zone i

$H(M)^h_{pi}$ is the number of type h mobile households initially located in zone i

$F(M)^H_{ti}$ is the initial space occupied by mobile households

Therefore, if no notable change has occurred in the zone, the model locates the new households close to the existing households who share the same typology or have similar household attributes. It can also be deduced that the households already established in the zone will tend to stay in the same location unless they suffer some form of considerable change in their preferences or in the utility of their current location.

The change in the utility of a location is calculated by

$$\Delta V^h_{pi} = \theta^{Uh}_p \cdot \Delta U^h_{\Delta t,i} + \theta^{Ah}_p \cdot \Delta A^h_{\Delta t,i} + \theta^{Qh}_p \cdot \Delta Q^h_{\Delta t,i} + \theta^{Rh}_p \cdot \Delta R^h_{\Delta t,i} \quad (9.8)$$

where:

$\Delta U^h_{\Delta t,i}$ is the change, during period Δt, in the housing consumption utility of type h located in zone i

$\Delta A^h_{\Delta t,i}$ is the change, during period Δt, in the accessibility of zone i for type h households

$\Delta Q^h_{\Delta t,i}$ is the change, during period Δt, in the quality of housing in zone i

$\Delta R^h_{\Delta t,i}$ is the change, during period Δt, in the environmental quality perceived by type h housing in zone i

$\theta^{Uh}_p, \theta^{Ah}_p, \theta^{Qh}_p, \theta^{Rh}_p$ are the parameters to be estimated

The location process is analysed in the following way. First, the housing consumption utility is calculated for each home in each zone. This provides the space that each household would occupy in each zone and, in turn, allows us to calculate the change in utility for each household locating in each zone. All the households are then located, and a comparison is made with the number of available homes. If availability is surpassed, the rents of each zone are adjusted and the process is repeated. The number of available homes can be modified using models that determine a proportion of the homes that have to remain vacant. The model can also be applied considering variable demand.

9.2.3   MUSSA AND MUSSA II MODELS

The MUSSA model has been designed to predict the location of agents, residents and companies in urban areas. It was originally presented in Martinez (1996) and later improved by Martínez and Donoso (2001).

The model is based on market equilibrium. Location is established through a process of bids, where it is assumed that each dwelling is acquired by the highest bid. It is assumed that all the agents will be located, and therefore the supply will satisfy all the demand.

The fundamental difference that was added by the new version, the MUSSA II model, consists of the inclusion of logit models to establish the supply of available real estate. As Logit models are also used for the demand, equilibrium is reached by solving equations based as a group on logit-type choice models.

For modelling purposes, the real estate is differentiated by type ($v \in V$), by location ($i \in I$) and by a vector of attributes ($z = (z_{vik}, k \in K)$). The users are classified into groups ($h \in H$) according to socioeconomic variables. The suppliers of real estate are also classified into groups ($j \in J$) according to differences in production costs. All the users are considered to be rational, and any idiosyncratic differences are addressed by the stochastic part of the logit model.

The MUSSA model considers each property to be a unique product that is dependent on a series of specific variables such as its location. Therefore, the bidding process is able to extract the willingness to pay for the users in a similar way to the theoretical assumptions of the Alonso location model (1964).

9.2.4   THE URBANSIM MODEL

The UrbanSim model was designed at the end of the 1990s to support decision-making in the field of planning, transport analysis and land use on a metropolitan urban scale (Waddell 2000, 2002, Waddell et al. 2003). A series of elements were defined to develop the model that have been maintained to the present day:

* *Representation of individual agents*: Initially households and companies, but then later people and jobs in a disaggregated way.
* *Representation of the supply, the characterisation of land and the development of real estate, on a suitable scale*: Initially, a mixture of plots and zones, later, cells of a size defined by the user.

- The adoption of a dynamic system with year long time cycles.
- The use of the real estate market as the main engine of the model with the supply and demand implicitly included, along with the resulting property prices.
- The use of standardised discrete choice models to represent the decisions of households and companies. The most commonly used model has been MNL.
- Integration of the existing transport systems in the study area in order to calculate the accessibility of each property and its influence on location choices.
- Adoption of an open-source license for developing the software, currently based on Python, and distribution on the web since 1998 through www. urbansim.org.

Since its implementation and practical application to the metropolitan zone of Eugene–Springfield (Waddell 1998, 2000, Waddell et al. 2002), many of the elements mentioned about UrbanSim have been adopted by other models. Figure 9.3 shows a flow diagram following the UrbanSim model. Note how the model is conditioned by external macro-economic models that estimate the patterns of economic evolution. It is also worth mentioning that the relationship with trip models works in both directions.

Residential location is performed in stages in the UrbanSim model, as can be seen in Figure 9.4.

The process starts by using a population synthesiser to generate a series of households for the base year being simulated, thereby creating a virtual population with similar characteristics to the real population (Beckman et al. 1996). Each household is represented in a database with a unique household indicator, and all its characteristics such as number of members, income and number of workers. Finally, a unique location identifier is also established.

During the first stage, a demographic transition model establishes the number of new households that are going to locate in a given zone. The relocation model also establishes the number of households that wish to move house. In both the cases, the household's location identifier becomes null. During the second stage, the model selects the required demand for housing by choosing all the households without a location identifier. The new home is chosen by applying a choice model to each household based on the utility calculated from the characteristics of the alternative houses and the interactions between the characteristics of housing and households. In practice, the most commonly used model is MNL (see Chapter 6). This procedure provides the calculation of the choice probability of one dwelling compared to another. Finally, once the probabilities have been obtained, one of the algorithms provided by the UrbanSim model assigns a home to each household until the demand has been met.

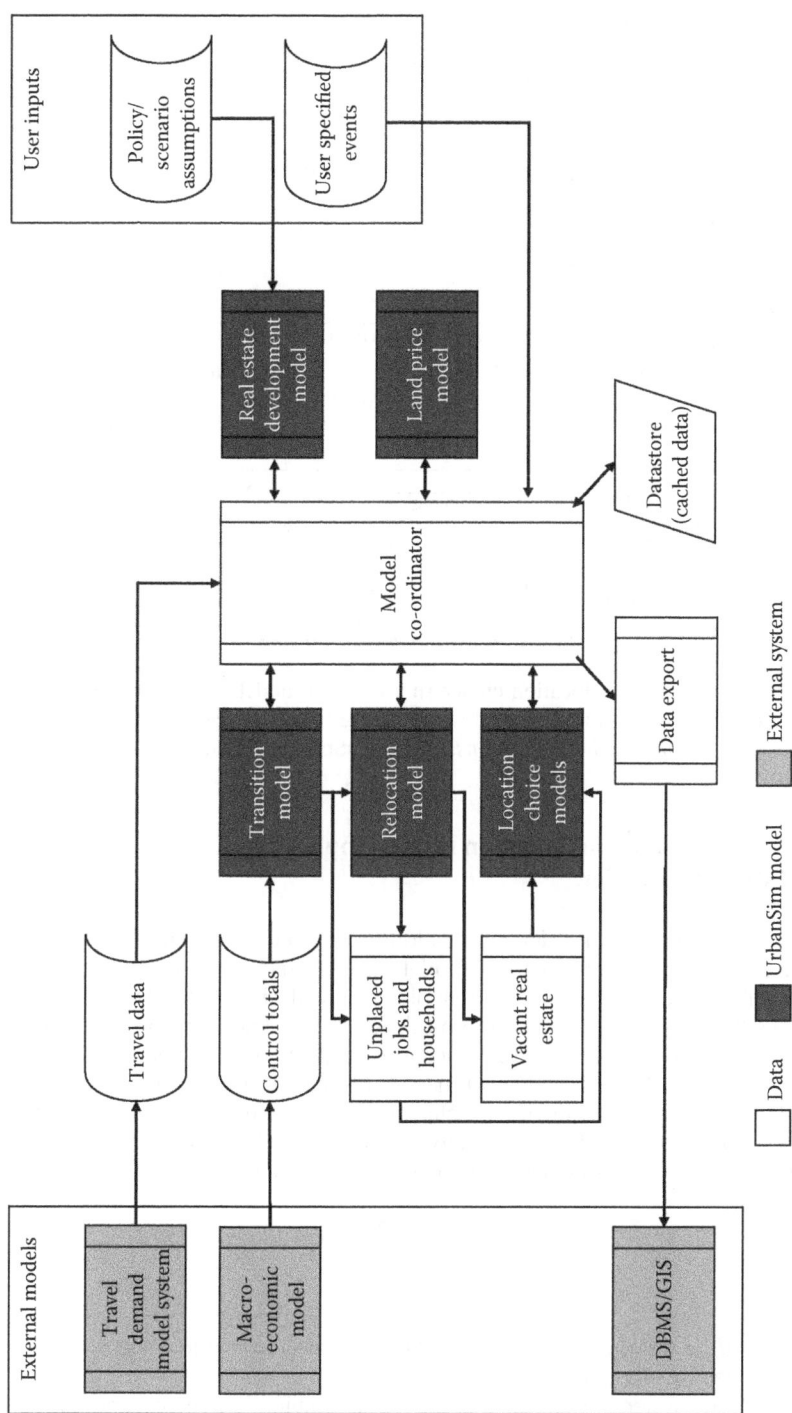

FIGURE 9.3   Flow diagram showing the structure of the UrbanSim model. (Based on Waddell, P., *J. Am. Plann. Assoc.*, 68, 297–343, 2002.)

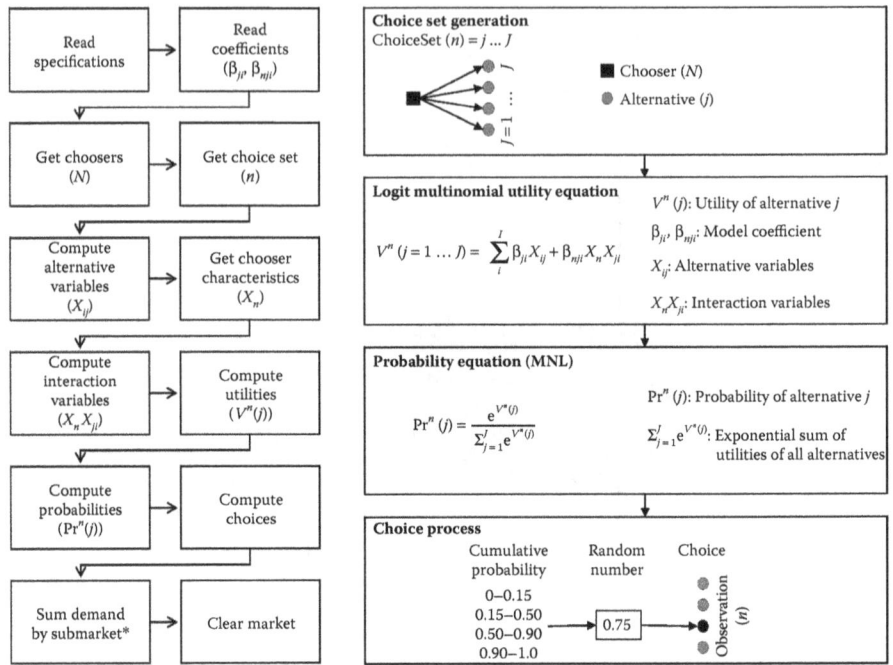

FIGURE 9.4 The process of location choice in a location model. *A submarket is group-ing of alternatives by location and property type. (Based on Pagliara, F. et al., *Residential Location Choice: Models and Applications*, Springer Science & Business Media, Berlin, Germany, 2010.)

Example 9.1: Residential Location Model Applied to the City of Santander

This example describes a residential location model that is estimated using data from the urban area of Santander. The model is based on hypotheses derived from random utility theory, in other words, individuals will choose the location that maximises their utility. Given that the modeller cannot know in absolute terms how individuals will value different locations, a probabilistic type of discrete choice model is postulated. The consumers of residential space will evaluate the different zones according to environmental attributes, distance/travel time to work and other characteristics. The probability that worker i chooses zone o as their place of residence conditioned to working in zone d is given by

$$P^i_{\text{res-cond}}(o|d) = \exp[V^i(o|d)] \Big/ \sum_o \exp[V^i(o|d)] \tag{9.9}$$

where:
$P^i_{\text{res-cond}}(o|d)$ is the probability that worker i chooses to live in zone o conditioned by working in zone d
$V^i(o|d)$ is the systematic utility given to worker i living in zone o conditioned
by working in zone d

The model is thereby specified with a MNL formulation (de Dios Ortuzar 2000). Different individuals can be grouped according to income level (medium–low and high, for example) and different models specified for each class. Furthermore, given the large number of alternatives, a random sample can be taken in order to reduce the size of the database and the need for greater resources to estimate the models. If the random sample taken from the alternatives is large enough, it will guarantee that the parameters obtained will be unbiased (McFadden 1977) (see also Chapter 6). In this case, five alternatives were sampled from the 42 that were available (see Chapter 8) for all the individuals. The use of this technique, however, impedes the estimation of specific constants in the utility functions.

Residential location can be made to depend on variables that are related to transport (journey times or overall costs), the location of activities (employment distribution) or real estate prices, which can be endogenous to the LUTI model. Residential location can also depend on external data such as the number of dwellings in each zone.

In this example, the systematic utility of location in zone o, given that the place of employment of an individual is in zone d, has been specified as being dependent on the following:

- $C(o,d)$ is the cost of travelling from home to work for each employee i (segmented by income level, medium–low or high).
- $P(o)$ is the average price of housing in zone o.
- $S(o)$ is the housing stock in zone o.
- Pres(o) is a dummy variable indicating if the zone is prestigious (city centre zone, environmentally desirable areas etc.).

The segmentation of the sample into two income levels allows different variables to be considered in each of the utility functions, and, at the same time, the aggregation error is reduced. This is caused by the parameters being forced to be identical for each sub-sample group, which is then, therefore, assumed to be homogenous in terms of tastes. Different criteria can be used to perform the segmentation; in this case, all the individuals with incomes greater than 2500 Euros per month were considered as a high-income group, whereas those with lower incomes were grouped in a category referred to as medium–low income households. Finally, the utility functions have been specified, following Equations 9.13 and 9.14 for high-income and medium–low-income households, respectively:

$$V_{(o|d)}^{i} = \beta_{i,C} \cdot C(o,d) + \beta_{i,S} \cdot S(o) + \beta_{i,\text{Pres}} \cdot \text{Pres}(o) \qquad (9.10)$$

$$V_{(o|d)}^{i} = \beta_{i,C} \cdot C(o,d) + \beta_{i,P} \cdot P(o) + \beta_{i,S} \cdot S(o) \qquad (9.11)$$

The parameters of the utility function should be estimated using maximum likelihood and specific choice software (NLOGIT or BIOGEME are two of the best known software packages). Furthermore, as can be seen in Equation 9.10, the location of the high-income individuals depends on the home–work journey cost, the available housing stock in each zone and a prestige variable, which takes a value of 1 if the zone has particularly attractive characteristics and enjoys a clearly positive social standing. The medium–low income households (Equation 9.11) value the home–work journey cost, the average house price in the zone and the

TABLE 9.1

Location Model Parameters Estimated According to Income Levels

Variable	Residential Location – High Income			Residential Location – Medium–Low Income		
	β	t	Sig.	β	t	Sig.
CT	−0.1311	−2.202	0.028	−0.1061	−3.336	0.001
VI	0.8674	−2.310	0.021	1.0983	5.830	0.000
PG	0.3285	1.671	0.095	−	−	−
PE	−	−	−	−1.5414	−4.812	0.000
$L(\theta)$	−235.035			−776.415		
$L(0)$	−495.230			−1677.91		
R^2	0.52			0.53		

Notes: CT is the home–work journey cost.
VI is the natural logarithm of the number of houses in the zone.
PG is the dummy variable referring to the prestige of the zone.
PE is the natural logarithm of the average price per m² of the homes in the zone.

available housing stock in each location. These specifications can be adapted to any study area and to the availability of data in order to improve the model's fit.

The results of estimating the parameters of the systematic utility functions are shown in Table 9.1.

The signs of the estimated parameters allow the modeller to check if the variables are having the desired effect. For high-income households, in the case of residential location, it was only the parameter of the CT variable that resulted negative. Which is theoretically consistent as the model is based on the hypothesis that residents will tend to prefer locations closer to their place of work. The estimation software also provides a test to check if each parameter is statistically different from zero. As a general rule, if the value of the test is above 1.96, the parameter is different from zero at a 95% confidence level. The log-likelihood of the model provides its goodness of fit with the data and can take values between 0 (perfect fit) and −∞. In this example, the fit is located around −235,035. Turning to the results of estimating the parameters for households with medium–low income levels, all the variables had a correct sign and were significant at a 95% confidence level. Note that, in agreement with the theory, the parameter of the PE variable had a negative sign, indicating that for medium–low income households, the average square meter price of housing clearly implies disutility.

Once the parameters of the utility functions have been estimated, the probabilistic location models can be calculated. If the study area is, hypothetically, considered to be closed from the point of view of the labour market, the hypothesis that the supply of jobs is taken up by the internal demand of employees can be added. By applying this hypothesis, it can be deduced that the number of workers $w^i(o)$ of type i locating in o is equal to

$$w^i(o) = \sum_d P^i_{\text{res-cond}}(o \mid d) \cdot \text{Emp}^i(d) \qquad (9.12)$$

where $Emp^i(d)$ is the total number of jobs in zone d for socioeconomic class i. Therefore, the total number of jobs by employee socioeconomic class needs to be found for all the zones in the study area. This data can be obtained by applying the following expression:

$$Emp^i(d) = \sum_a h_a^i(d) \cdot Emp_a(d) \qquad (9.13)$$

where:
    $Emp^a(d)$ is the number of jobs belonging to economic sector a in zone d
    $h_a^i(d)$ is a factor representing the ratio of type i workers employed in an activity
        belonging to sector a in zone d

Finally, given the number of workers present in each zone, it is possible to find the total number of residents present by using a coefficient $k(o)$, which represents the ratio between residents and workers in each zone o:

$$res(o) = k(o) \cdot \sum_i \sum_d P_{res\text{-}cond}^i(o \mid d) \cdot Emp^i(d) \qquad (9.14)$$

With this final expression and starting from the proposed hypothesis, it is feasible to calculate the number of residents locating in each zone. The number of residents has also been shown to depend on the number of jobs existing in zone d, which leads to the configuration of an equilibrium problem between the location of residents and economic activities that should be addressed in the combined model (see Chapter 13).

9.3 ECONOMIC ACTIVITY LOCATION MODELS

Economic activity location models have been developed using a wide range of methodologies and theories, suggesting the complexity associated with modelling this phenomenon. A brief state of the art is presented in the following to show the different methods that have been used and their historic evolution. Highlights among these methods are as follows:

- Input–output analysis using matricial methods
- Computational models of general equilibrium (CGE)
- Discrete choice through maximum utility, typically represented by logit models
- Application of the theories of maximum utility in micro-simulation models
- Models based on agents or cellular automata

The key to modelling this phenomenon lies in identifying the engines driving the economic development of an area. It needs to be pointed out that the methodologies mentioned earlier are not exclusive, and they are frequently combined to create models that better represent reality.

In the Lowry (1964), the location of activities that is dependent on local demand is performed using a gravity model that is dependent on the location of residents.

As was the case with the residential location models, the activity models were later refined based on entropy maximisation theory. The MEPLAN (Echenique and Hargreaves 2003) software was developed at the beginning of the 1980s, combining input–output tables, to simulate the workings of the economic system (Leontief 1966), as well as random utility theory, to simulate location choices. The TRANUS (de la Barra 1989) and PECAS (Hunt and Abraham 2003) models followed the methodology proposed by MEPLAN; however, each of them developed specific aspects related to the location of activities. A summary of various activity location models is presented in Sections 9.3.1 and 9.3.2, all of which are currently being used and represent the state of the art. Because some activity location models such as DELTA (see Section 9.2.2) follow the same methodology as residential location models, it has been decided to show a series of alternative models in order to complete the overall vision of the current state of the art in the field of location choice modelling.

9.3.1 IRPUD MODEL

The IRPUD model is considered to be an interregional mobility and location simulation model for a metropolitan area (Wegener 1982, 1983, 1985, 1996, 2011). The study area is divided into zones connected by the existing transport supply. Time is modelled in periods of one or more years, depending on the required level of detail.

The complete model is based on a modular structure (see Figure 9.5), which uses the different modules and sub-models to estimate the location of activities, real estate

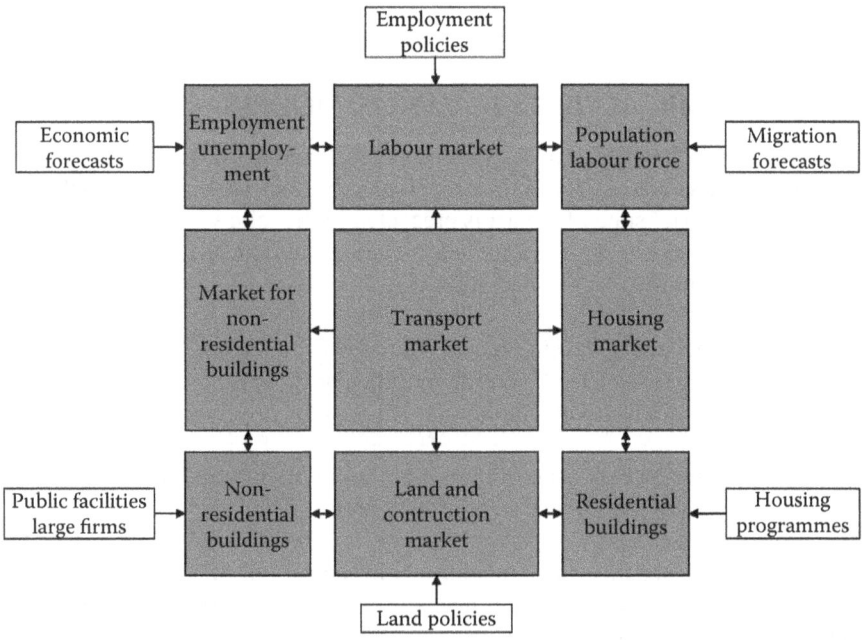

FIGURE 9.5 The IRPUD model. (Based on Wegener, M., The IRPUD model, *Spiekermann & Wegener in Dortmund*, http://www.spiekermann-wegener.com/mod/pdf/AP_1101_IRPUD_ Model.pdf, 2011.)

developers and houses. The model consists of the following six sub-models, although only those relating to activities are developed in this section:

- *The transport sub-model*: Calculates the journeys made by the model's users for the modes of on-foot/cycling, public transport and car.
- *The aging sub-model*: Makes the changes caused by the passage of time required by the model; changes are based on stochastic Markov-type updating models without being addressed as choice problems.
- *The public programmes sub-model*: This sub-model applies all the public policies, which the modeller wants to include and which may affect any environment of the model.
- *The private construction sub-model*: Considers the investment and the location of new investment in infrastructure, from both construction companies and real estate promoters.
- *The labour market sub-model*: Decision model where the workers decide to stay in their current job or change to another within the zone being studied.
- *The household sub-model*: Simulates the intra-zonal migration of households.

The IRPUD activity location model is not based around a single sub-model, rather it is a process where various of them interact. This location can in turn be divided into two sections: on the one hand, modelling the location of commercial buildings and industries and on the other hand, the location of the workers in their respective homes.

In both sections, the ageing and public programme sub-models establish the temporal evolution criteria and the possible modifications fixed externally by the modeller. By using choice models, the construction sub-model establishes where the commercial and industrial firms are located and where the workers' homes are located or relocated, thereby providing the zonal location of activities. By considering all the changes established in the previous stages, the labour market sub-model defines the employee decisions to change jobs or not. This decision finally affects the housing sub-model that defines if the homes associated with the workers change location or not.

The probabilities associated with the choice models applied in the sub-models correspond to discrete choice models that seek to maximise the utility for individuals by using a logit type of formulation.

9.3.2 MEPLAN Model

The MEPLAN model is based on a spatially segregated input–output matrix (Echenique and Hargreaves 2003). In the MEPLAN model, the aggregated predictions about employment and its spatial location are separated, so the locations of housing and services, along with their interactions, are also separated.

The general LUTI model works sequentially, on the one hand, providing an overall aggregated estimation of the employment and population in a region, then disaggregate into different types of employment, using the association of the available population to each type of employment. This process is organised by using the input–output structure. The generated employment is later located by using a location model.

The employment location model works in two stages. The first stage calculates the available land, considering a linear relationship proportional to some, yet undetermined, independent variables (Equation 9.18). Once the available land is established, the employment is generated by making predictions for each sector and each zone. This second stage also follows a linear relationship (Equation 9.19). These two linear models are estimated using ordinary least squares.

$$y_i' = a_0 + \sum_g a_g x_{ig} \tag{9.15}$$

$$z_i' = b_0 + \sum_g b_g v_{ig} + \vartheta y_i' \tag{9.16}$$

where the independent variables x_i and v_i are associated with the parameters a_g and b_g, respectively, and the parameter ϑ represents the weight of the available space in the first model when estimating the number of jobs z_i'.

These two models need to consider a series of constraints, where the most important being that the number of jobs estimated for each zone cannot exceed the total number of jobs calculated using the input–output model.

Example 9.2: Economic Activity Location Model for the City of Santander

The economic activity location model presented as a practical application to the city of Santander is based on the following expression:

$$\text{Emp}_a(d) = P_a(d) \cdot \text{EMP}_a \tag{9.17}$$

where:
  $\text{Emp}_a(d)$ is the number of jobs located in zone d belonging to a economic sector a
  $P_a(d)$ is the probability of a type a job being located in zone d
  EMP_a is the total number of type a jobs in the study area

To simulate the location decisions of the activities, in consistency with the residential location model presented earlier, the modeller turns to random utility theory, in order to model the location decisions as discrete choices. It is postulated that the private agents (companies) assign a utility to each zone and choose the one that maximises it. The utility is once again assumed to be composed of two parts: a systematic part $V_a(d)$ and a random part ε_a. If these residuals are identically and independently distributed Gumbel, then the probability $P_a(d)$ of locating the activity in zone d is given by

$$P_a(d) = \exp[V_a(d)] \Big/ \sum_d \exp[V_a(d')] \tag{9.18}$$

The systematic utility $V_a(d)$ has been specified as a linear combination of the following attributes:

$A^{pas}(d)$ is an accessibility indicator of zone d for the population. In this case, it is calculated using a gravity-type indicator (see Chapter 3).
$Res(d)$ is the number of residents present in zone d.
$E^{bas}(d)$ is the base employment (see below) present in zone d.
$Centre(d)$ is a dummy variable that takes a value of 1 if the zone belongs to the central area of the urban system, where a large number of commercial premises and services are located.

The model can differentiate between activities, which in urban modelling research are generally classified into four categories (Nuzzolo and Coppola 2005):

- The basic sector, which is generally dependent on exporting to outside the system being studied, so its location is not directly dependent on the distribution of the internal demand (population and other activities).
- Activities aimed at the internal demand, such as retail and non-specialised services, which depend on the location of the demand, that is, the population and other activities.
- Representative activities, such as those that locate, depending on attractive zonal characteristics due to prestige or centrality.
- Activities with a low spatial efficiency, in other words, activities that generally require a lot of land to function, such as car dealerships and heavy industry.

The proposed activity location model will only consider those activities in which their location depends on the distribution of the population, in other words, those activities aimed at the internal demand. The parameters associated to retail and services are normally considered and estimated separately. This reduces aggregation errors by allowing different parameters to be estimated for each sector, as with residential location, where people were separated into different income groups. The systematic utilities are calculated using the following expression:

$$V_a(d) = \beta_{a,acc} \cdot A^{pas}(d) + \beta_{a,Res} \cdot Res(d) + \beta_{a,E} \cdot E^{bas}(d) + \beta_{a,cen} \cdot Centre(d) \qquad (9.19)$$

The location of economic activities may depend on accessibility, influenced by the characteristics of the transport infrastructure, and on residents present in each zone, in other words, the residential location. Basic sector jobs group together all those activities considered in the classification presented earlier, as not being dependent on the distribution of the internal demand, making them part of the input data for the model.

The results of the location model are presented in Table 9.2.

In the case of the commercial retail location model, the parameters CEN, ACC and EMP were significant at a 95% confidence level. All the presented parameters had a positive sign that made them consistent with the opening theoretical hypotheses. Certain zonal constants were also specified to highlight which larger area each zone belonged to. These constants allowed the fit of the model to be improved and made up for the lack of specific constants, which could not be estimated because

TABLE 9.2

Parameters Estimated in the Location Models

Variable	Retail Location			Location of Services		
	β	t	Sig.	β	t	Sig.
CEN	1.683	12.372	0.000	1.7934	20.241	0.000
ACC	0.6868	7.279	0.000	0.1033	1.677	0.936
EMP	0.5177	9.799	0.000	0.4025	10.801	0.000
K2	2.4213	18.405	0.000	−0.2970	−2.964	0.003
K3	−0.3676	−2.334	0.020	−0.4994	−4.457	0.000
K4	−0.9482	−4.145	0.000	–	–	–
$L(\theta)$	−1237.5			−2578.57		
$L(0)$	−4208.6			−7475.33		
R^2	0.70			0.65		

Notes: CEN is a dummy variable signalling if the zone is part of the urban centre.
ACC is the accessibility of the zone.
EMP is the number of basic sector jobs in the zone.
K2, K3 and K4 are zonal constants.

the model was calibrated using sampling of alternatives. The study area was divided into four large sub-areas to allow a better replication of the starting location pattern.

In the case of the activity location model for the service sector, the results showed that all the variables had the correct sign and were significant at 95% except in the case of the ACC variable, which was only 90% significant. In spite of that, it was decided to keep it in the model, given its theoretical importance. The variable centre was once again the most important when providing location utility to the agents.

After estimating the parameters of the utility functions, the choice probabilities can be calculated for each activity along with the distribution of jobs according to economic sector by using expression (9.17).

9.4 CONCLUSION

This chapter has shown the variety of residential and activity location models that are currently available in the literature. The evolution and use of the different modelling methodologies have led to the opportunity of choosing one type of model or another according to the availability of data and the requirements of the modeller.

Although the methodologies applied for residential and activity locations are based on a diversity of attributes or characteristics, they share a certain common theoretical foundation based on the idea of maximising utility. Furthermore, location models do not function in an isolated way, rather they form part of wider modelling systems conforming LUTI models.

The practical examples developed in this chapter have shown how residential and activity location models can be applied by reviewing the consistency of their results in real cases.

REFERENCES

Alonso, W. 1964. *Location and Land Use: Toward a General Theory of Land Rent.* Publications of the Joint Center for Urban Studies of the Massachusetts Institute of Technology and Harvard University. Cambridge, MA: Harvard University Press.

Anas, A. 1983. Discrete choice theory, information theory and the multinomial logit and gravity models. *Transportation Research Part B: Methodological* 17 (1): 13–23. doi:10.1016/0191-2615(83)90023-1.

Beckman, R. J., K. A. Baggerly, and M. D. McKay. 1996. Creating synthetic baseline populations. *Transportation Research Part A: Policy and Practice* 30 (6): 415–429.

Bossard, J. H. S. 1932. Residential propinquity as a factor in marriage selection. *American Journal of Sociology* 38: 219–224.

de Dios Ortuzar, J. (Ed.) 2000. *Modelos econométricos de elección discreta.* Santiago, Chile: Universidad Católica de Chile ed.

de la Barra, T. 1989. *Integrated Land Use and Transport Modelling: Decision Chains and Hierarchies, Cambridge Urban and Architectural Studies.* Cambridge, UK: Cambridge University Press.

Echenique, M. H., and A. J. Hargreaves. 2003. *Cambridge Futures 2: What Transport for Cambridge.* Cambridge, UK: Cambridge Futures/University of Cambridge.

Hansen, W. G. 1959. How accessibility shapes land use. *Journal of the American Institute of Planners* 25 (2): 73–76.

Hunt, J. D., and J. E. Abraham. 2003. Design and application of the PECAS land use modelling system. *8th International Conference on Computers in Urban Planning and Urban Management*, Sendai, Japan.

Ikle, F. C. 1954. Sociological relationship of traffic to population and distance. *Traffic Quarterly* 8 (2): 123–136.

Leontief, W. 1966. *Input-Output Economics.* New York: Oxford University Press.

Lill, E. 1891. *Das Reisegesetz und seine Anwendung auf den Eisenbahnverkehr.* Wien, Austria: Spielhagen & Schurich.

Lowry, I. S. 1964. *A Model of Metropolis, Memorandum.* Santa Monica, CA: Rand Corporation.

Martinez, F. 1996. MUSSA: Land use model for Santiago city. *Transportation Research Record* 1552: 126–134.

Martínez, F., and P. Donoso. 2001. Modeling land use planning effects: Zone regulations and subsidies. In *Travel Behaviour Research: The Leading Edge*, D. Hensher (Ed.), pp. 647–658. Amsterdam, the Netherlands: Pergamon.

McFadden, D. L. 1977. *Modelling the Choice of Residential Location.* New Haven, CT: Cowles Foundation for Research in Economics, Yale University.

Nuzzolo, A., and P. Coppola. 2005. S.T.I.T.: A system of mathematical models for the simulation of land-use and transport interactions. *Proceedings of European Transportation Conference*, Strasbourg, France.

Oryani, K., and B. Harris. 1996. Enhancement of DVRPC's travel simulation models - Review of land use models and recommended model for DVRPC. In *Land Use Compendium*. Washington, DC: U.S. Department of Transportation. Federal Highway Administration.

Pagliara, F., J. Preston, and D. Simmonds. 2010. *Residential Location Choice: Models and Applications.* Berlin, Germany: Springer Science & Business Media.

Putman, S. H. 1979. *Urban Residential Location Models, Studies in Applied Regional Science v. 13.* Boston, MA: Martinus Nijhoff.

Ravenstein, E. G. 1885. The laws of migration. *Journal of the Royal Statistical Society* 48:167–235.

Reilly, W. J. 1931. *The Law of Retail Gravitation.* New York: W.J. Reilly.

Senior, M. L., and A. G. Wilson. 1974. Explorations and syntheses of linear programming and spatial interaction models of residential location. *Geographical Analysis* 6 (3): 209–238.

Simmonds, D., and O. Feldman. 2005. Land-use modelling with DELTA: Update and experience. *Proceedings of the Ninth International Conference on Computers in Urban Planning and Urban Management (CUPUM)*. http://www.cix.co.uk/~davidsimmonds/main/pdfs/cupum_paper_354.pdf.

Simmonds, D., and A. Skinner. 2003. *The South and West Yorkshire Strategic Land-Use/Transportation Model*. Chichester, UK: Wiley.

Simmonds, D. C. 1999. The design of the DELTA land-use modelling package. *Environment and Planning B: Planning and Design* 26 (5): 665–684.

von Thünen, J. H. 1826. *Der isolierte staat in beziehung auf landwirtschaft und nationaloekonomie*. Jena. Translated by C. M. Wartenburg (1966). The Isolated State. Oxford, UK: Oxford University Press.

Waddell, P. 1998. Oregon prototype metropolitan land use model. *Proceedings of the Conference on Transportation, Land Use, and Air Quality*, Portland, OR.

Waddell, P. 2000. A behavioral simulation model for metropolitan policy analysis and planning: Residential location and housing market components of UrbanSim. *Environment and Planning B: Planning and Design* 27 (2): 247–263.

Waddell, P. 2002. UrbanSim: Modeling urban development for land use, transportation, and environmental planning. *Journal of the American Planning Association* 68 (3): 297–343.

Waddell, P., A. Borning, M. Noth, N. Freier, M. Becke, and G. Ulfarsson. 2003. Microsimulation of urban development and location choices: Design and implementation of UrbanSim. *Networks and Spatial Economics* 3 (1): 43–67.

Waddell, P., M. Outwater, C. Bhat, and L. Blain. 2002. Design of an integrated land use and activity-based travel model system for the Puget Sound region. *Transportation Research Record* 1805: 108–118.

Wegener, M. 1982. *Aspects of Urban Decline: Experiments with a Multilevel Economic-Demographic Model for the Dortmund Region*. Laxenburg, Austria: IIASA.

Wegener, M. 1983. *Description of the Dortmund Region Model*. Dortmund, Germany: Institut für Raumplanung.

Wegener, M. 1985. The Dortmund housing market model: A Monte Carlo simulation of a regional housing market. In *Microeconomic Models of Housing Markets*, pp. 144–191. Berlin, Germany: Springer.

Wegener, M. 1996. Reduction of CO2 emissions of transport by reorganisation of urban activities. In *Transport, Land-Use and the Environment*, pp. 103–124. Berlin, Germany: Springer.

Wegener, M. 2011. The IRPUD model. *Spiekermann & Wegener in Dortmund*. http://www.spiekermann-wegener.com/mod/pdf/AP_1101_IRPUD_Model.pdf.

Wilson, A. G. 1970. *Entropy in Urban and Regional Modelling, Monographs in Spatial and Environmental Systems Analysis 1*. London, UK: Pion.

Young, E. C. 1924. *The Movement of Farm Population*. Vol. 426. Cornell University Agricultural Experiment Station.

Zipf, G. K. 1949. *Human Behavior and the Principle of Least Effort*. Cambridge, MA: Addison-Wesley.

10 Models for Simulating the Impact of Accessibility on Real Estate Prices

Rubén Cordera, Ángel Ibeas and Luigi dell'Olio

CONTENTS

This chapter introduces models that are designed to simulate the impact that access to opportunities has on both residential and commercial property prices. This type of model tries to answer questions such as follows: do changes in an area's accessibility cause positive or negative changes in the prices of nearby property? How relevant are these changes? Can some kind of mechanism be used to capture the added value generated in such a way that greater accessibility will result in further investment in transport and particularly in public transport?

10.1 INTRODUCTION TO THE IMPACTS OF ACCESSIBILITY ON REAL ESTATE VALUES

The possible impacts of a new transport infrastructure or policy on the surrounding area need to be considered when evaluating a project. Traditional cost-benefit assessments have normally only considered improvements in mobility through indicators such as time savings weighted by the value of user time as a measure of social benefit (Willumsen 2014). However, the introduction of a new transport infrastructure can generate indirect effects that result from improved accessibility to opportunities and their capitalisation represented in the surrounding property prices. The existence of this type of effect has a theoretical basis provided by the so-called Henry George theory (1884). According to this theory, in a spatial economy where the concentration of economic activity is based around local public institutions and when the size of the population is optimal, the aggregated value of the land rents is equal to expenditure on public services (Arnott and Stiglitz 1979). So, according to this theory, a single tax payment for all the land rents would ideally be enough to pay for the public services provided in the area (Peddle and Gaffney 2009). This result has even been corroborated under less strict suppositions than sustained by the original Arnott and Stiglitz theorem (Behrens et al. 2010).

However, the empirical measurement of these benefits is a complex task because of the multiple effects that new transport infrastructure or services can have on the surrounding area. In spite of their variety, these benefits can be classified into three large groups (Fogarty et al. 2008):

- *Environmental benefits*: Less traffic congestion, lower fuel consumption, improved air quality, less urban sprawl and others
- *Fiscal benefits*: Lower expenditure on roads and parking spaces, higher property prices, greater productivity in the economy
- *Social benefits*: Improved social cohesion, greater equality, better public health, fewer accidents and increased access to opportunities

Nevertheless, the impacts of public services in general and transport in particular do not necessarily have to be beneficial, which complicates how to measure their diversity and how to differentiate their effects. An example of evaluating public transport whilst considering a diversity of effects that are only slightly addressed by conventional cost-benefit analysis is provided by Lewis and Williams (1999). These authors evaluated several positive effects of public transport such as greater equality and its provision of access for people without a private vehicle (the elderly, children, low-income earners), its greater efficiency along popular corridors and its relationship with high density urban development and diversity of uses. Lewis and Williams estimated these benefits for the United States as being in the region of 45–60 billion US dollars, well over the public transport budget of around 20 billion dollars. Studies of this type underline the importance of considering the indirect positive effects of public services such as transport. Disgracefully, there is no standard methodology to estimate the magnitude of each benefit, and empirical studies have normally concentrated on trying to measure some of them in particular whilst trying to control the presence of other factors as much as possible. In this sense, measuring the benefits generated in property prices is particularly useful as it can reflect many of the indirect benefits that result from improved accessibility and transport.

Some of the calculations about the impact that changes in transport have on property prices can be seen in Tables 10.1 and 10.2 for the cases of North America and Europe, respectively. Debrezion et al. (2007) performed a meta-analysis using a sample of 50 studies to detect an average increase of 4.2% in property values as a result of being located at least 400 m from a train station. Most of the research to date has

TABLE 10.1
Impacts of Transport on Housing Prices in North America

Study Area	Type of Transport	Detected Effect	Source
Toronto (Canada)	Metro	+20% close to stations	Bajic (1983)
Miami (United States)	Metro	Positive but small	Gatzlaff and Smith (1993)
Boston (United States)	Local train	+6.7% (settlements with station)/−20% (122 m from the track)	Armstrong Jr (1994)
San Francisco (United States)	Train	+10% to +15% (up to 400 m from stations)	Cevero (1996)
Chicago (United States)	Metro	+20% (300 m from station)	Gruen (1997)
Atlanta (United States)	Metro	−19% (400 m from station)/+3.5% (1600–4800 m from stations)	Bowes and Ihlanfeldt (2001)
Santa Clara County (United States)	Metro and Light rail	+45% (up to 400 m from stations)	Cervero and Duncan (2002a)
Dallas (United States)	Light rail	+24.7% (up to 400 m from stations)	Clower and Weinstein (2002)
Los Angeles (United States)	Metro, train, Light rail and BRT	Unequal and inconsistent effects	Cervero and Ducan (2002)
St. Louis (United States)	Light rail	+15.5% (up to 850 m to stations in some models)	Garret and Castelazo (2004)
Portland (United States)	Light rail	+10%	Hass-Klau and Crampton (2005)
Eastern Massachusetts (United States)	Commuter rail	+9.6 to +10.1% (Boroughs with a station)	Armstrong and Rodríguez (2006)
Buffalo (United States)	Light rail	+2 to +5% close to stations	Hess and Almeida (2007)
Washington DC (United States)	Metro	+2.5% (every 160 m towards a station). Only rentals	John and Sirmans (2009)
Ottawa (Canada)	Rail transit	+5.33$ per additional metre closer to a station	Hewitt and Hewitt (2012)

Source: Fogarty, N.A. et al., *Capturing the Value of Transit*, Center for Transit-Oriented Development, Federal Transit Administration, Oakland, CA, 2008; Correia, G. and Viegas, J.M., *J. Simul.*, 3, 61–68, 2009; Smith, J.J. and Gihring, T.A., *Financing Transit Systems Through Value Capture. An Annotated Bibliography*, Victoria Transport Policy Institute, Victoria, Canada, 2006.

TABLE 10.2
Impacts of Transport on Housing Prices in Europe

Study Area	Type of Transport	Detected Effect	Source
Tyne and Wear (United Kingdom)	Metro	+1.7%	Pickett (1984)
Manchester (United Kingdom)	Light rail	Without effects	Forrest et al. (1996)
Sheffield (United Kingdom)	Light rail	+1.7%	Dabinett (1998)
London (United Kingdom)	Metro (Jubilee Line)	Without effects	Jubilee Line Extension Impact Study Unit (2000)
Newcastle (United Kingdom)	Light rail	+20%	Hass-Klau and Crampton (2005)
Fribourg (Germany)	Light rail	+3% (Rental)	Hass-Klau and Crampton (2005)
Lisbon (Portugal)	Metro	+3.5% to +5.2% (Up to 10 min walk to the nearest stop)	Martínez and Viegas (2009)
Madrid (Spain)	Metro	+2.2% to 3.2% (1000 m to metro stop)	Dorantes et al. (2011)
Santander (Spain)	Private vehicle to the town centre	+1.1% to +1.8% (per additional minute of closeness to the centre)	Ibeas et al. (2012)
Santander (Spain)	Train	−2.7% to −6% (up to 500 m from stations)	Ibeas et al. (2012)

Source: Fogarty, N.A. et al., *Capturing the Value of Transit*, Center for Transit-Oriented Development, Federal Transit Administration, Oakland, CA, 2008; Correia, G. and Viegas, J.M., *J. Simul.*, 3, 61–68, 2009; Smith, J.J., and Gihring, T.A., *Financing Transit Systems Through Value Capture. An Annotated Bibliography*, Victoria Transport Policy Institute, Victoria, Canada, 2006.

concentrated on measuring the impact of distance from public transport installations such as metro, light rail, train or bus stations. Other studies have also assessed the effects of accessibility measured using indicators such as those presented in Chapter 3 (Ibeas et al. 2012).

Generally, almost all research has shown the existence of positive effects that the closer you are to public transport stations or the greater accessibility to opportunities is, the higher the property prices are. However, in some cases and especially in the case of railway stations, negative effects have also been detected derived from externalities such as noise, pollution and movement of passengers in the surrounding area. These negative cases were detected in research carried out in Boston and Atlanta (United States), whereas only minor or even null effects were detected in studies carried out in Manchester and London (United Kingdom).

There is less research on the impact of transport on commercial property prices. However, the results have generally been very similar to those of residential properties showing the greater price increases (Cervero and Duncan 2002b) for those premises located closest to transport installations (Debrezion et al. 2007).

10.2 VALUE CAPTURE POLICIES

Given the evidence that improved accessibility to transport facilities and therefore to a greater diversity of opportunities can be capitalised on by property owners, proposals have been made in the specialised academic literature and in the operational domain to design procedures that retain this added value for the public. The aim of these policies is to stop the property owners, especially with concentrated ownership, from making excessive profits from public services and investments. The policies that aimed at capturing this added value are able to recuperate part of these benefits by obtaining extra funds for further investment in infrastructure or public transport services (Fogarty et al. 2008).

Value capture policies can therefore be defined as a public financing mechanism based on the recuperation of all or part of the added value gained by the property owners from the creation of new public installations or services. Although the property owners in the areas immediately surrounding the public investments will be particularly benefitted by them, the value capture policies look to redistribute these profits among a much wider spread of the population.

On a practical level, value capture policies have been introduced in various urban areas using a great variety of mechanisms. One of the first to be proposed was in the United States after the introduction of new public transport systems during the 1970s in cities such as Washington, DC, San Francisco or Atlanta. Hagman and Miscyznski (1978) proposed the introduction of taxes based on the resulting increases in land values for the affected properties.

Fogarty et al. (2008) classified value capture policies for public transport investment into four basic groups:

- *Special tax*: It is a specific tax limited in time aimed at the real estate benefitting from the project. This is the case in special zones that are set up in the United States such as Portland, Tampa or Atlanta.
- *Added value tax*: It is a tax that captures all the increases in the values of the properties affected by the project.
- *Joint development*: It is based on the development of new transport projects with financing extracted from new real estate developments nearby. This policy is related to the transit-oriented development (TOD) in which the highest density urban development occurs in areas with greater accessibility, close to locations such as metro or tram stations. Emblematic examples of this type of joint development can be found in Hong Kong and Singapore (Chi-Man Hui et al. 2004), although they have also been applied in different urban areas in the United States such as Washington or Philadelphia (Zhao et al. 2012). Sometimes, this type of development is based on direct selling or on permit

concessions for the rights to develop real estate in the affected areas, whereas in other cases permits are granted for private housing developments, or bonds are issued for high-density developments as in New York (Cervero et al. 2004).

- *Real estate development fee*: It is a special tax directed at new property developments resulting from the extension of public services, including transport. An example of this is the transit impact development fee introduced in San Francisco with variable tariffs per square metre depending on land use.

10.3 HEDONIC MODELS FOR CALCULATING THE IMPACTS OF ACCESSIBILITY ON REAL ESTATE VALUES

The specialised literature provides examples of two methodologies to evaluate the effects of different variables on real estate prices. First is the repeated sales technique, based on a linear regression. This regression is performed between the differences of at least two sales prices for the same property at different times and a series of dummy variables representing the chosen time intervals. This type of model was proposed by Bailey et al. (1963), and it continues to be used today as it does not require information about the characteristics of each property sold if the characteristics have not changed significantly between sales. However, this method has the disadvantage of depending on repetitive sales, a kind of data that is difficult to obtain in many study areas.

The most commonly used alternative method based on cross-sectional data is hedonic regression. With hedonic regression, a heterogeneous product z, for example a property, is split into different characteristics or attributes z_n

$$Z = (z_1, z_2, z_3, ... z_n) \qquad (10.1)$$

The characteristics z_n can be classified into three main types: environmental/social, structural characteristics and transport/accessibility to opportunities. This kind of model has been used in studies over many years (Court 1939) even though it was Rosen (1974) who formulised the supply and demand equilibrium model for heterogeneous goods. This chapter will only refer to the hedonic model estimated during a first phase, in other words, the implicit prices model based on market equilibrium, that is able to provide the willingness to pay for marginal increases in any of the specified attributes. Estimating the second phase of a hedonic model also allows the researcher to calculate the structural parameters of the supply and demand curves for the market of the heterogeneous product (Malpezzi 2008). Therefore, the hedonic function becomes conceptualised as the envelope function of the supply and demand functions of producers and consumers, respectively. The hedonic pricing function on its own does not reveal information about the preferences of the producers and consumers that generated it. Nevertheless, McFadden (2013) has highlighted that under certain conditions, the preferences of the consumers can be recovered directly from the hedonic function. These conditions are as follows: (1) consider that all consumers have identical preferences and (2) consider that all consumers have identical perceptions about the unobservable attributes. Under these hypotheses, the market prices are determined independently from the market and production structures, so the parameters of the hedonic regression directly represent the consumers' preferences.

Although there is no consensus in the literature about the functional form nor about the variables to be specified in the model, a common way of specifying the hedonic pricing model is as follows:

$$\ln(P) = X\beta + \varepsilon \tag{10.2}$$

where:

P is the vector with the property sale price

X are a series of independent variables corresponding to the different characteristics of the heterogeneous product

β is a vector of a parameter to be estimated

ε is a vector of independent and identically distributed errors between observations

If the hedonic model is specified as Equation 10.2, then the estimated parameters can be directly interpreted as semi-elasticities, in other words, the percentage change in the dependent variable per unit change in the independent variable. Furthermore, as the model is linear, it can be directly estimated using ordinary least squares. Nevertheless, hedonic models have received their share of criticism due to various weaknesses. First, they require large trustworthy databases of property prices, and these are sometimes difficult to obtain, create and update. Second, they are prone to errors derived from the important spatial effects that are normally present in real estate data, for example, the presence of sub-markets in certain zones with their own specific equilibriums (spatial heterogeneity) or from the effects of diffusion between the prices of nearby properties (spatial autocorrelation). Spatial hedonic models were developed to control this type of effects using techniques derived from spatial econometrics (Anselin 1988, 2010). Third, the lack of a theoretical guide means that the specification of the models could be erroneous and certain relevant variables could be omitted (Sirmans et al. 2005). However, all these problems can be minimised through researching the available theoretical and empirical literature and by correctly evaluating compliance with the suppositions of the proposed hedonic model.

Hedonic pricing methodology has been applied in various cases within the context of land use–transport interaction (LUTI) models. The UrbanSim model developed by the Urban Planning Department and the Computer Science Department of the University of Washington (Waddell et al. 2003) incorporates a land pricing sub-model through a hedonic regression that is included in the following equation:

$$P_{ilt} = \alpha + \delta\left(\frac{V_i^s - V_{it}^c}{V_i^s}\right) + \beta X_{ilt} \tag{10.3}$$

where:

P_{ilt} is the price of type i property in location l at time t

V_{it}^c is the rate of vacancy of property type i at time t

V_i^s is the long term structural vacancy rate

X_{ilt} is a vector of the property's structural attributes and characteristics of the surroundings

α, β and δ are parameters to be estimated

This methodology allows the hedonic model within UrbanSim to capture both the relative prices of each one of the specified characteristics as the structural effects derived from the property vacancy rate in each location. The intercept of the hedonic regression is fitted at annual time intervals as a function of the current property vacancy rate and the long-term structural rate of vacant properties. If the current number of vacant properties rises, then the regression intercept drops, whereas if the number of vacant properties drops, the intercept will correct itself by rising to increase property prices.

Coppola and Nuzzolo (2011) also proposed a similar real estate pricing model as part of a LUTI model to evaluate the effects of different transport policies in the urban area of Naples (Italy). The model was formulated on a zonal level as

$$X_j(o) = \gamma_0 X_{oj}(o) + \gamma_1 \left(\frac{\sum_i \delta_{ij} A^i(o)}{S_j(o)} \right)^{\gamma_2} + \varepsilon \tag{10.4}$$

where:

$X_j(o)$ is the final hedonic price of the type j properties in zone o
$X_{oj}(o)$ is the initial hedonic price dependent on the characteristics of the zone
$A^i(o)$ is the location demand of a type i agent in zone o
δ_{ij} is an index indicating whether the type j property is demanded by type i agents
$S_j(o)$ is the supply of locations in zone o for property type j
γ_0, γ_1, γ_2 are parameters to be estimated
ε is an error term

Therefore, the price determined by the initial hedonic regression is modified according to demand A^i and location supply S_j specified for each zone.

Example 10.1: Hedonic model for residential property in the city of Santander

One of the sub-models required in the estimation of the LUTI model in Santander is the property price simulator. Similarly to Equation 10.4, the proposed model is based on a hedonic regression that also considers location demand to modify the property prices in each zone. The model does not, therefore, provide a complete simulation of market equilibrium as other models described in the following section do, but it does allow predictions to be made, given changes in the interaction between land use and transport.

The zoning presented below is as described in Chapter 8, where the urban area of Santander and its sphere of influence was divided into a total of 42 zones. The data required to estimate a model of this type are generally aggregated on a zonal level or disaggregated on an individual property level. If the availability of data allows it, it is always better to work with disaggregated data on an individual property level as this can always be aggregated later if the model requires it. In this case, disaggregated data were available to estimate the hedonic pricing models. The following variables were available in the database:

- P: Price of the property.
- IMPROV: The property requires major improvement or not.
- ROOMS: Number of rooms in the property.
- TER: Presence or absence of balcony/terrace.
- GAR: Presence or absence of at least one garage.
- LIFT: Presence or absence of a lift.
- POPSQM: Ratio between the residential population in the zone and the number of built residential square metres. This variable makes it possible to set the price as a function of the supply and demand of housing in each zone.
- EMP: Number of jobs located in the zone of the property.
- ACCA: Gravity-type accessibility indicator (see Chapter 3) in each zone. It is calculated by considering the total journey time on public transport (on-board journey time, waiting time and transfer time, if present) and the employment opportunities present in each of the destination zones.
- CBD: Relative type of accessibility indicator considering the total journey time on public transport to arrive at the central urban area.
- TRANS: Presence or absence of a public transport stop within 400m from the property.
- CEN: Represents being part of the city's urban centre. This dummy variable should be based on knowledge of the study area and the limits considered defining the central zone, given the presence of greater administrative, business, commercial activity and so on.
- COM: Presence of the property in a commercial area. As in the previous case, this dummy variable should be fixed as 1 in those properties that, according to the modeller or some kind of quantitative indicator, form part of a commercial zone.
- GREEN: Presence of green spaces in the zone where the property is located.
- PG: Particularly prestigious zone. This dummy variable has a more qualitative nature allowing the analyst to capture the effects of special prestige, generally provided by an environmental externality, which cannot be captured by the other variables.

The descriptive statistics, together with the measurement units of each of the variables, are shown in Table 10.3. Figure 10.1 represents the distribution of property prices in the study area. The highest prices are found in the neighbourhoods to the east and north east of the city centre and a north–south price gradient can also be seen from the coast to the interior.

The S-1 and S-2 hedonic models estimated using ordinary least squares are presented in Table 10.4. Two models were estimated because the ACCA and CBD accessibility variables were strongly correlated and could not be specified at the same time. The S-1 model only had the ACCA variable in its specification, whereas the S-2 had the CBD. In both cases, the dependent variable of the model has been the logarithmically transformed price of the properties. This means that the parameters estimated by the different variables can be interpreted as semi-elasticities.

TABLE 10.3
Descriptive Statistics of the Variables Contained in the Property Database
(N = 1562 Properties)

Variable	Description	Units	Mean	Standard Deviation	Minimum	Maximum
P	Property price	Euros	321,425	264,433	60,000	3,000,000
IMPROV	The property requires major improvement	1/0	0.07	0.26	0	1
ROOMS	Number or rooms	–	2.97	1.15	0	12
TER	Presence of balcony/ terrace	1/0	0.26	0.44	0	1
GAR	Presence of garage	1/0	0.56	0.50	0	1
LIFT	Presence of LIFT	1/0	0.52	0.50	0	1
POPSQM	Population/ Built square metres	Persons per m²	0.04	0.01	0.02	0.06
EMP	Jobs	N° of Jobs	2,598.86	2,718.94	441	11,359
ACCA	Active Accessibility using public transport	–	0.85	0.91	0.04	3.56
CBD	Journey time to centre using public transport	Minutes	52.14	21.57	13.45	82.24
TRANS	Public transport stop within 400m from the property	1/0	0.89	0.31	0	1
CEN	Part of the central zone	1/0	0.07	0.26	0	1
COM	Part of a commercial zone	1/0	0.14	0.35	0	1
GREEN	Presence of green spaces	1/0	0.21	0.41	0	1
PG	Zone of special prestige	1/0	0.15	0.36	0	1

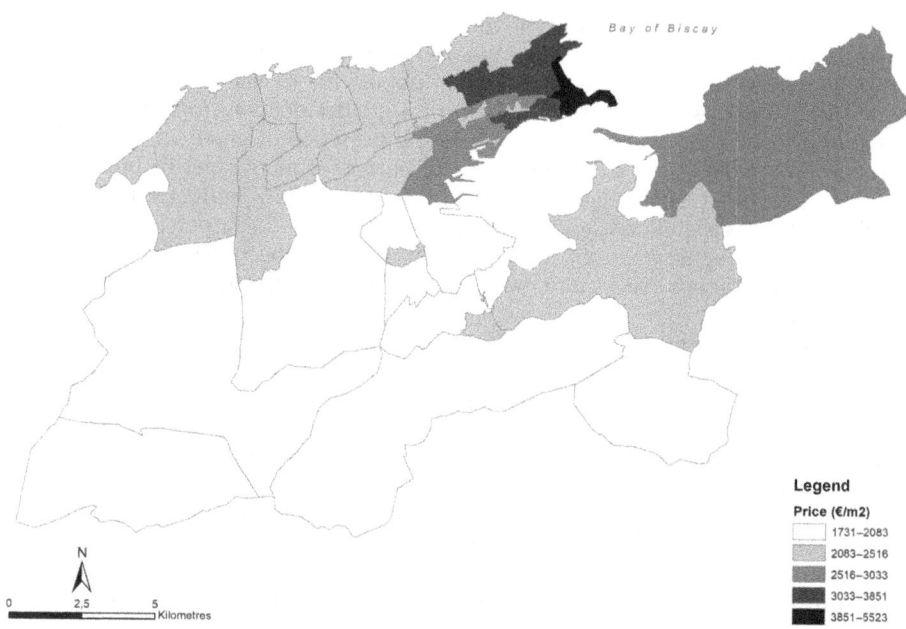

FIGURE 10.1   Housing prices (euros/m²) in the city of Santander.

TABLE 10.4
Estimated Hedonic Pricing Models

Variable	S-1	S-2
(Intercept)	12.384 (.000)	12.460 (.000)
IMPROV	–0.122 (.001)	–0.127 (.000)
ROOMS	0.282 (.000)	0.285 (.000)
TER	0.065 (.002)	0.069 (.001)
GAR	0.251 (.000)	0.278 (.000)
LIFT	0.171 (.000)	0.162 (.000)
POPSQM	–49.35 (.000)	–41.39 (.000)
POPSQM²	521.14 (.000)	440.87 (.001)
EMP	0.000 (.651)	0.000 (.897)
ACCA	0.055 (.000)	–
CBD	–	–0.004 (.000)
TRANS	0.025 (.438)	–0.09 (.791)
CEN	–0.128 (.128)	–0.147 (.076)
COM	–0.069 (.056)	–0.032 (.379)
GREEN	0.224 (.000)	0.247 (.000)
PG	0.274 (.000)	0.205 (.000)
R^2	0.61	0.62
R^2 adjusted	0.61	0.62
Test F	174.43 (.000)	179.33 (.000)

One aspect to highlight about the models is that the POPSQM variable has also been introduced squared. This squared specification allows the variable's negative effects on prices due to the higher population density to be differentiated from the positive effect derived from a greater demand for residential location. Environmental variables such as the presence of green spaces and structural variables such as the presence of balconies or terraces in housing are also shown to have a positive effect. The accessibility indicator relative to the CBD was significant and had the correct sign (−0.4% per additional minute of journey time to the urban centre), whereas the integrated gravity accessibility indicator ACCA had the expected positive sign and it was clearly significant.

Equations 10.5 and 10.6 also bring together the final specifications of both models. It needs to be pointed out that the specification of the hedonic models should comply with the hypothesis of a conventional regression model (Gujarati and Porter, 2009). Of particular relevance is the need to avoid specification errors due to the omission of relevant variables as the model could show biased parameters or situations of high co-linearity between independent variables that would impede differentiating their effects on the dependent variable. Furthermore, the presence of spatial dependence in the residuals of the model could also generate biased or inefficient parameters. This problem can be addressed using spatial econometrics techniques as explained in Chapter 12.

$$
\begin{aligned}
\ln(P_i) = {} & \beta_1 + \beta_2 IMPROV_i + \beta_3\, ROOMS_i + \beta_4 TER_i + \beta_5 GAR_i \\
& + \beta_6 LIFT_i + \beta_7 POBSQM_i + \beta_8 POBSQM^2{}_i + \beta_9 EMP_i \\
& + \beta_{10} ACCA_i + \beta_{11} TRANS_i + \beta_{12} CEN_i + \beta_{13} COM_i \\
& + \beta_{14} GREEN_i + \beta_{15} PG_i + \varepsilon_i
\end{aligned}
\tag{10.5}
$$

$$
\begin{aligned}
\ln(P) = {} & \beta_1 + \beta_2 IMPROV_i + \beta_3 ROOMS_i + \beta_4 TER_i + \beta_5 GAR_i \\
& + \beta_6 LIFT_i + \beta_7 POBSQM_i + \beta_8 POBSQM^2{}_i + \beta_9 EMP_i \\
& + \beta_{10} CBD\_TP_i + \beta_{11} TRANS_i + \beta_{12} CEN_i + \beta_{13} COM_i \\
& + \beta_{14} GREEN_i + \beta_{15} PG_i + \varepsilon_i
\end{aligned}
\tag{10.6}
$$

10.4   MODELS BASED ON SIMULATING THE PROPERTY MARKET TO CALCULATE THE IMPACTS OF ACCESSIBILITY

Different LUTI models have been designed to directly simulate the land market, either in its residential sector or its commercial sector, or in both sub-sectors. These economic models were mainly developed from the 1970s and 1980s as practical implementations of urban economic theory modelling the real estate market.

One of the first models of this type was developed by Anas and collaborators. The Chicago Area Transportation – Land Use Analysis System (CATLAS) model was developed at the start of the 1980s (Anas 1987) to evaluate the effects of improvements made to the transport systems on the real estate market in the urban area of Chicago in terms of both supply and demand. More recently, MUSSA (Modelo de Usos del Suelo de Santiago, Santiago land use model) model developed by Francisco J. Martinez and his team for the Chilean government (Martinez 1992, Martinez and Henriquez 2007) has also combined the simulation of the real estate market with the discrete choice techniques developed from the works of Domencich and McFadden (1975). In the MUSSA model, the proportion of consumers H of type h in a location i looking for a property of type v is given by:

$$P_{hvi} = \frac{H_h \exp(\mathrm{WP}_{hvi})}{\sum_g H_g \exp(\mathrm{WP}_{gvi})} \qquad (10.7)$$

As can be seen, Equation 10.7 is a multinomial logit type of expression (see Chapter 6) in which the choice does not only depend on systematic utility but also on the random residuals distributed Gumbel independently and identically between location alternatives. WP is willingness to pay, which can be specified taking into account the characteristics z of the property, the characteristics of the surroundings and the socioeconomic characteristics of the consumer looking for a property:

$$\mathrm{WP}_{hvi} = \mathrm{WP}_h(d, z, \beta_h, y_h, U_h) + \varepsilon_h \qquad (10.8)$$

where:
  y_k is the income of the user
  U_h is the maximum level of utility reachable
  β_h are parameters to be estimated
  ε_k is a random residual

The WP should therefore be estimated using a sample of agents (households or companies) that should be available with all the variables that need to be specified about the characteristics of the property and the consumer.

The expected market price for location i of property type v is equal to the maximum bid expected from the possible buyers:

$$p_{vi} = \left(\frac{1}{\mu}\right) \log\left[\sum_g H_g \exp(\mathrm{WP}_{gvi})\right] \qquad (10.9)$$

On the supply side, the property developers maximise their profits calculated as the price (p_{vi}) minus the development costs (c_{vi}) by also using a logit type of formula. The proportion of real estate developers R for a housing type v in a location i is equal to

$$R_{vi} = \frac{\exp(p_{vi} - c_{vi})}{\sum_{vi} \exp(p_{vi} - c_{vi})} \qquad (10.10)$$

The model calculates the market equilibrium as the point in which the total number of consumers equals the number of available properties. The transactions could be subject to constraints derived from the regulation of land use.

REFERENCES

Anas, A. 1987. *Modeling in Urban and Regional Economics, Fundamentals of Pure and Applied Economics*. Chur, Switzerland: Harwood Academic Publishers.

Anselin, L. 1988. *Spatial Econometrics: Methods and Models, Studies in Operational Regional Science 4*. Dordrecht, the Netherlands: Kluwer Academic Publishers.

Anselin, L. 2010. Thirty years of spatial econometrics. *Papers in Regional Science* 89 (1): 3–25. doi:10.1111/j.1435-5957.2010.00279.x.

Armstrong, Jr., R. J. 1994. Impacts of commuter rail service as reflected in single-family residential property values. *Transportation Research Record* 1466: 88–97.

Armstrong, R., and D. Rodríguez. 2006. An evaluation of the accessibility benefits of commuter rail in Eastern Massachusetts using spatial hedonic price functions. *Transportation* 33 (1): 21–43. doi:10.1007/s11116-005-0949-x.

Arnott, R. J., and J. E. Stiglitz. 1979. Aggregate land rents, expenditure on public goods, and optimal city size. *The Quarterly Journal of Economics* 93: 471–500.

Bailey, M. J., R. F. Muth, and H. O. Nourse. 1963. A regression method for real estate price index construction. *Journal of the American Statistical Association* 58 (304): 933–942.

Bajic, V. 1983. The effects of a new subway line on housing prices in metropolitan Toronto. *Urban Studies* 20 (2): 147–158.

Behrens, K., Y. Kanemoto, and Y. Murata. 2010. The Henry George Theorem in a second-best world. *Journal of Urban Economics* 85: 34–51.

Bowes, D. R., and K. R. Ihlanfeldt. 2001. Identifying the impacts of rail transit stations on residential property values. *Journal of Urban Economics* 50 (1): 1–25. doi:10.1006/juec.2001.2214.

Cervero, R., and M. Ducan. 2002a. *Land Value Impacts of Rail Transit Services in Los Angeles County*. Washington, DC: National Association of Realtors, Urban Land Institute.

Cervero, R., and M. Duncan. 2002b. Benefits of proximity to rail on housing markets: Experiences in Santa Clara County. *Journal of Public Transportation* 5 (1): 1–18.

Cervero, R., and M. Duncan. 2002c. Transit's value-added effects: Light and commuter rail services and commercial land values. *Transportation Research Record: Journal of the Transportation Research Board* 1805: 8–15.

Cervero, R., National Research Council (U.S.). Transportation Research Board, Transit Cooperative Research Program, United States Federal Transit Administration, and Transit Development Corporation. 2004. *Transit-oriented Development in the United States: Experiences, Challenges, and Prospects, TCRP Report*. Washington, DC: Transportation Research Board.

Cevero, R. 1996. Transit-based housing in the San Francisco Bay area: Market profiles and rent premiums. *Transportation Quarterly* 50 (3): 33–49.

Chi-Man Hui, E., V. S.-M. Ho, and D. K.-H. Ho. 2004. Land value capture mechanisms in Hong Kong and Singapore: A comparative analysis. *Journal of Property Investment & Finance* 22 (1): 76–100.

Clower, T. L., and B. L. Weinstein. 2002. The impact of Dallas (Texas) area rapid transit light rail stations on taxable property valuations. *TheAustralasian Journal of Regional Studies* 8 (3): 389–400.

Coppola, P., and A. Nuzzolo. 2011. Changing accessibility, dwelling price and the spatial distribution of socio-economic activities. *Research in Transportation Economics* 31 (1): 63–71. doi:10.1016/j.retrec.2010.11.009.

Correia, G., and J. M. Viegas. 2009. A conceptual model for carpooling systems simulation. *Journal of Simulation* 3 (1): 61–68.

Court, A. T. 1939. Hedonic price indexes with automobile examples. In *The Dynamics of Automobile Demand*, Horner, S.L. (Ed.), pp. 98–119. New York: General Motors.

Dabinett, G. 1998. Realising regeneration benefits from urban infrastructure investment: Lessons from Sheffield in the 1990s. *Town Planning Review* 69 (2): 171–189.

Debrezion, G., E. Pels, and P. Rietveld. 2007. The impact of railway stations on residential and commercial property value: A meta-analysis. *The Journal of Real Estate Finance and Economics* 35 (2): 161–180. doi:10.1007/s11146-007-9032-z.

Domencich, T. A., and D. McFadden. 1975. *Urban Travel Demand: A Behavioral Analysis, Contributions to Economic Analysis 93*. Amsterdam, the Netherlands: North-Holland.

Dorantes, L., A. Paez, and J. Vassallo. 2011. Analysis of house prices to assess economic impacts of new public transport infrastructure: Madrid Metro Line 12. *Transportation Research Record: Journal of the Transportation Research Board* 2245: 131–139.

Fogarty, N., N. Eaton, D. Belzer, and G. Ohland. 2008. *Capturing the Value of Transit*. Oakland, CA: Center for Transit Oriented Development, Federal Transit Administration.

Forrest, D., J. Glen, and R. Ward. 1996. The impact of a light rail system on the structure of house prices: A hedonic longitudinal study. *Journal of Transport Economics and Policy* 30 (1): 15–29.

Garret, T. A., and M. D. Castelazo. 2004. *Light Rail Transit in America: Policy Issues and Prospects for Economic Development*. St. Louis, MO: Federal Reserve Bank of St. Louis.

Gatzlaff, D. H., and M. T. Smith. 1993. The impact of the Miami metrorail on the value of residences near station locations. *Land Economics* 69 (1): 54–66.

George, H. 1884. *Progress and Poverty: An Inquiry into the Cause of Industrial Depressions, and of Increase of Want with Increase of Wealth, the Remedy*. London, UK: W. Reeves.

Gruen, A. 1997. *The effect of CTA and METRA stations on residential property values*. Report to the Regional Transportation Authority.

Gujarati, D. N., and D. C. Porter. 2009. *Basic Econometrics*, 5th ed. Boston, MA: McGraw-Hill Irwin.

Hagman, D., and D. Miscyznski. 1978. *Windfalls for Wipeouts: Land Value Capture and Compensation*. Chicago, IL: American Society of Planning Officials.

Hass-Klau, C., and G. Crampton. 2005. Economic impact of light rail investments: Summary of the results for 15 urban areas in France, Germany, UK and North America. In *Urban Transport Development: A Complex Issue*, Gunilla, J. and Tengström, E. (Eds.), pp. 245–255. Berlin, Germany: Springer.

Hess, D. B., and T. M. Almeida. 2007. Impact of proximity to light rail rapid transit on station-area property values in Buffalo, New York. *Urban Studies* 44 (5–6): 1041–1068. doi:10.1080/00420980701256005.

Hewitt, C. M., and W. E. Hewitt. 2012. The effect of proximity to urban rail on housing prices in Ottawa. *Journal of Public Transportation* 15 (4): 3.

Ibeas, A., R. Cordera, L. dell'Olio, P. Coppola, and A. Dominguez. 2012. Modelling transport and real-estate values interactions in urban systems. *Journal of Transport Geography* 24: 370–382. doi:10.1016/j.jtrangeo.2012.04.012.

John, B., and S. Sirmans. 2009. Mass transportation, apartment rent and property values. *Journal of Real Estate Research* 12 (1): 1–8.

Jubilee Line Extension Impact Study Unit. 2000. *Property Market Study*. Working Paper No. 32. London, UK: University of Westminster.

Lewis, D., and F. L. Williams. 1999. *Policy and Planning as Public Choice: Mass Transit in the United States*. Vol. 5, Aldershot, UK: Ashgate.

Malpezzi, S. 2008. Hedonic pricing models: A selective and applied review. In *Housing Economics and Public Policy*, K. Gibb and T. O'Sullivan (Eds.), pp. 67–89. Oxford, UK: Blackwell.

Martinez, F. J. 1992. The bid-choice land-use model: An integrated economic framework. *Environment & Planning A* 24 (6): 871–885.

Martinez, F. J., and R. Henriquez. 2007. A random bidding and supply land use equilibrium model. *Transportation Research Part B: Methodological* 41 (6): 632–651.

Martínez, L., and J. Viegas. 2009. Effects of transportation accessibility on residential property values. *Transportation Research Record: Journal of the Transportation Research Board* 2115 (1): 127–137.

McFadden, D. L. 2013. *The New Science of Pleasure*. Cambridge, UK: National Bureau of Economic Research.

Peddle, F. K., and M. Gaffney. 2009. The hidden taxable capacity of land: Enough and to spare. *International Journal of Social Economics* 36 (4): 328–411.

Pickett, M. W. 1984. The effect of the Tyne and wear metro on residential property values. *Supplementary Report* 825. Berkshire, UK: Transport and Road Research Laboratory.

Rosen, S. 1974. Hedonic prices and implicit markets: Product differentiation in pure competition. *Journal of Political Economy* 82 (1): 34–55.

Sirmans, G. S., D. A. Macpherson, and E. N. Zietz. 2005. The composition of hedonic pricing models. *Journal of Real Estate Literature* 13 (1): 3–43.

Smith, J. J., and T. A. Gihring. 2006. *Financing Transit Systems Through Value Capture. An Annotated Bibliography*. Victoria, Canada: Victoria Transport Policy Institute.

Waddell, P., A. Borning, M. Noth, N. Freier, M. Becke, and G. Ulfarsson. 2003. Microsimulation of urban development and location choices: Design and implementation of UrbanSim. *Networks and Spatial Economics* 3 (1): 43–67.

Willumsen, L. G. 2014. *Better Traffic and Revenue Forecasting*. London, UK: Maida Vale Press.

Zhao, Z. J., K. V. Das, and K. Larson. 2012. *Joint Development as a Value Capture Strategy in Transportation Finance, 2012*. Transportation finance, value capture.

11 Models for Simulating the Transport System

Borja Alonso, Rubén Cordera and Ángel Ibeas

CONTENTS

This chapter is a generic overview of the different models and methodologies used to simulate transport systems in urban and suburban areas. As the aim of a land use–transport interaction (LUTI) model is to support planning, it should be complemented by a strategic approach to modelling land use characteristics and the consequences of locating activities, in other words, the population's need to move from their residences to different centres of activity and vice versa. The need to travel and the associated costs perceived by the users of the transport system will be what initiates the reciprocal influence of the transport system on the activities system.

This chapter will neither be addressing models aimed at managing and operating traffic and passenger flows nor those designed to evaluate their evolution over short time frames such as approaches based on dynamic or quasi-dynamic models; rather the models described in this chapter will be framed within trip-based models because those are the kind that have traditionally been used in LUTI modelling.

11.1 INTRODUCTION TO MOBILITY AND THE TRANSPORT MODEL

It has already been explained in Chapters 1 through 10 that *mobility* is a socio-economic phenomenon originating from the spatial and temporal distribution of diverse social and economic activities. These activities generate the need for people and goods to be able to move between different origins and destinations throughout a territory where they occur. This is why the function of the *transport system* supplies the *infrastructure of services*, which allows these people and goods to arrive on time at the correct destinations to fulfil their objectives: work, business, shopping, leisure, study and so on.

An interaction will exist between the people and/or goods that wish to move/be moved between specific points or zones in a region (demand) using the available infrastructure and transport services (supply). A determined planning of the supply will have repercussions on the behaviour and distribution of the demand; in turn, the resulting distribution of the demand will create new needs and would result in a redesign and adaptation of the transport supply. It is this process of design and adaptation of the supply to the requirements of the demand that needs to be calculated by *transport planning*. Of course, there is an added difficulty that the demand is highly dynamic in terms of temporal and spatial distribution as well the heterogeneity of the individuals, frequently subjected to stochastic processes, whereas the supply is rigid in terms of infrastructure with a very restricted variability in terms of transport services (e.g. restrictions of fleets, budgets and labour conditions for workers).

Correct planning can only occur when suitable instruments are available to the planner, which, especially from the latter third of the twentieth century, has been applied transport modelling. Therefore, the modelling of transport is not the planning of transport but a decision-making planning tool, though in certain situations it may take on a major role which nevertheless will not be definitive (de Dios Ortúzar and Willumsen 2011).

The classic transport model has traditionally been approached as a sequential decision-making process for individuals who are about to make their journeys: the individual takes the decision to make a journey, and this journey is the result of the requirement to satisfy a need and/or fulfil an objective: go to work, go shopping and so on. Based on this motivation, the individual chooses a destination for their journey, the mode of transport to be used and finally, the route to be followed. The representation of this sequence leads to the classic four-stage model: trip generation, spatial (or zonal) distribution, modal choice and route choice or assignment (Figure 11.1).

Previously, this approach had to be addressed from an aggregated viewpoint that largely simplified the combinations of possible journey origins and destinations. This led to the need to spatially aggregate the territory using the so-called transport zones. In transport models, the journeys are not addressed as points of origin and destination but are addressed as movements made from an origin zone to a

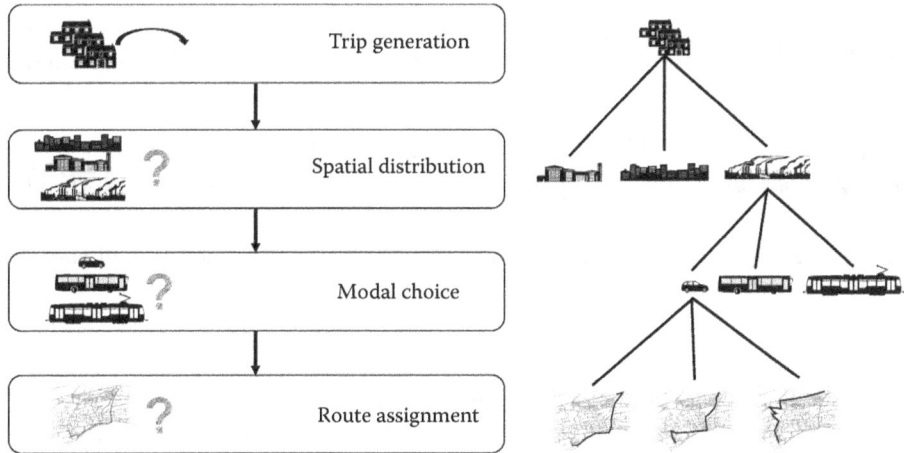

FIGURE 11.1   Structure of the four-stage classic transport model.

destination zone. As already stated, the zoning will be dependent on the system of activities and land use that are specific to each area of the territory or the region being modelled. Therefore the chosen zoning should be consistent with those used in the activities and land use model, as explained in Chapter 8.

In the following pages, the models that are more commonly used at each stage in the classic model will be explained within the context of a LUTI model.

11.2 TRIP GENERATION MODELS

11.2.1 BASIC CONCEPTS

Trip generation is a process that quantifies the journeys made by people residing in or those who are active in a determined urban area or those who use vehicles relating to that area (de Dios Ortúzar and Willumsen 2011). More specifically, a journey is a one-directional movement from a point (zone) of origin to a point (zone) of destination (Figure 11.2). With this simple description, the following categorisation is normally used:

- *Origin*: Point or zone where the journey starts.
- *Destination*: Point or zone where the journey ends.
- *Home-based journeys* (*HB*): These are journeys where one of its ends is the home of the person making the journey. It could be the origin or the destination.
- *Non-home-based journeys* (*NHB*): Where neither the origin nor the destination is the home.
- *Trip production*: Defined as the end household in a, HB journey or the origin of an NHB journey.
- *Trip attraction*: Defined as the non-household end of a, HB journey or the destination of an NHB journey.

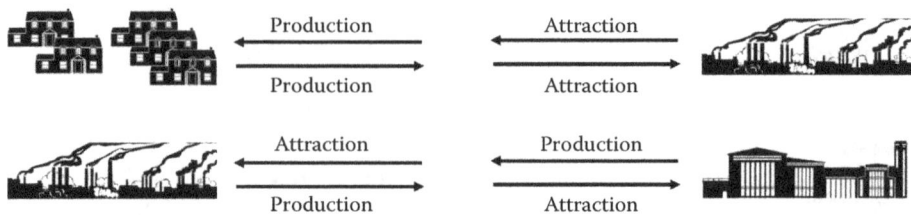

FIGURE 11.2 Trip production and attraction. (Based on de Dios Ortúzar, J. and Willumsen, L.G., *Modelling Transport*, John Wiley & Sons, Hoboken, NJ, 2011.)

Trip generation models are functional relationships between generated trips and explanatory variables. Knowing the values of the explanatory variables within a future perspective allows predictions to be made about future travel demand. The following three factors have historically been considered to affect travel demand:

- The location of activities and the characteristics of land use
- The socioeconomic characteristics of the residents in the area
- The extension, costs and quality of the available transport services

Apart from these factors, there is a fourth factor, *accessibility*, which, in turn, is conditioned by the other three and is especially relevant in LUTI models (see Chapter 3).

Logically, trip generation needs to be integrated into the overall model, so generation has to be estimated for each transport zone, in other words, a trip generation vector has to be obtained for zone *z*.

$$G_z = f\left(P_z, R_z, E_z, A_z, \dots\right) \qquad \forall z \qquad (11.1)$$

These vectors can be disaggregated according to journey typology (HB, NHB), journey purpose (study, work, leisure...), type of user (car owner, income level...), type of day (working, holiday, Saturday...) and, in the non-daily models, according to time period (morning, evening rush hour, midday, off peak...), as well as combinations of the above, to provide some examples. For example, there is clearly a relationship between trip purpose and time of day as the main reason behind the morning rush hour is when people go to work and/or study.

Figure 11.3 details the distribution of trip purpose in an origin–destination (O-D) survey (2009), thereby highlighting the following reasons: work at morning rush hour, later study and leisure/shopping during the intermediate periods of the day.

For all the study area, the number of trips produced should be equal to the number of attracted trips, although not necessarily for any one zone in particular. This is because, by definition, the HB trips are always produced by a zone containing the household and attracted by the zone containing the other end, in any direction in which the journey is made. Whereas the NHB journeys are produced by the origin zone and attracted by the destination zone.

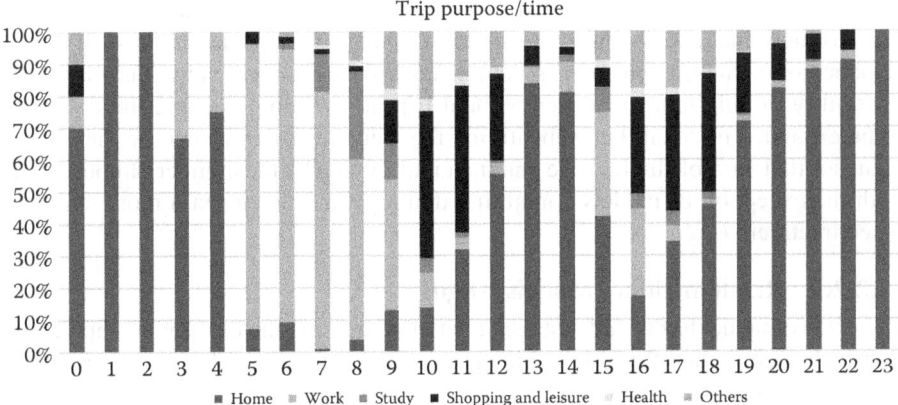

FIGURE 11.3 Distribution of journeys according to purpose and time of day.

Taking the earlier into account, it can be easily deduced that trip production will be greater than trip attraction in the zones that are mainly residential, whereas the attraction will be greater than the production in those zones that are mainly commercial, industrial or educational. This distinction between the production and attraction of journeys is important when using gravity models for the trip distribution phase as they consider the origin and the destination of journeys.

Therefore, the *trip generation stage* can be defined as that which aims to quantify and identify the journeys that terminate in the different zones dividing the study area.

11.2.2 METHODOLOGIES

11.2.2.1 Multiple Linear Regression

Due to its relative ease of implementation and conceptual simplicity, multiple linear regression (MLR) is the most frequently used tool in designing trip generation models. These models estimate the number of journeys generated by a determined zone (or household) as a linear function of certain variables related to the facilities and land use of the zone. The typical expression of an MLR model used to estimate trip generation (and attraction), either on an aggregated or disaggregated level, is as follows:

$$y_i^p = \beta_k^p + \sum_k \beta_k^p \cdot x_{ik}^p + \varepsilon_i^p \tag{11.2}$$

where:

$\varepsilon_i^p \sim N(0; \sigma^2)$

y_i^p is the number of trips of the type, purpose or category p generated by unit i. The unit can be an individual, a household or a zone.

β is the vector of parameters to be estimated (according to journey type)

x is the vector of explanatory variables (consistent with the unit of analysis)

ε_i^p is the error of estimation for unit i

This type of model can be applied to estimate trip production and attraction on a zonal or household scale. First, the zoning of the study area should be such that its sociodemographic composition is homogenous for each zone, notwithstanding the possibility of null zones or zones without trip information. Second, the data need to be expanded to zonal data substituting the value of each variable by the average value for that variable in the zone and then multiplying by the number of households, with the exception of models containing dummy variables when a more disaggregated treatment is required.

11.2.2.2   Random Utility Models (Logit)

This type of model has the advantage of better addressing the problem of trip generation inelasticity when faced with changes to infrastructure and zonal accessibility. These models first estimate the probability that each individual decides to make a trip and then multiply this probability by the number of individuals in that particular category. The expression of multinomial Logit (MNL, see Chapter 6) is normally used to determine this probability.

The utility function of an individual from class i making x journeys needs to be specified once again based on the socioeconomic variables of the individual, accessibility and so on. The choice group of the number of journeys will depend on the period of the analysis; so for 1 hour, only two alternatives are taken: to travel or not to travel ($x = 0, 1$) and the model is simplified by setting limits for longer periods ($x = 0, 1, 2, 3$ or more), leading to the following expression:

$$D_o^i[sh] = n_o(i)m_i(osh) = n_o(i)\sum_x xp^i(x/osh)$$

$$= n_o(i)\sum_x x\frac{\exp\left(\lambda V_{x/osh}^i\right)}{\sum_j \exp\left(\lambda V_{x/osh}^i\right)} \tag{11.3}$$

where:
  $D_o^i[sh]$ is the average number of trips originating in zone o for purpose s in period h by user category i
  $n_o(i)$ is the number of category i users in zone o
  $m_i(osh)$ is the average number of trips made by class i users, from zone o, for purpose s in period h
  $p^i(x/osh)$ is the probability of class i individual making x journeys, from zone o, for purpose s, in period h
  V^i is the utility function of making x journeys for class i individual from o for purpose s in period h

11.3   SPATIAL OR ZONAL DISTRIBUTION MODELS

11.3.1   Basic Concepts and Initial Considerations

It is stated in the introduction to this chapter that transport models addressed journeys as movements from an origin zone to a destination zone. Chapter 5 addressed spatial interaction models and explained that movement between pairs of zones can be represented by the O-D matrix. In this matrix, the sum of the trips for each of the rows should be equal to the total number of trips generated by the zone to which the row refers; similarly, the sum of the trips for each column should correspond to the number of trips attracted by the zone to which the column refers. Or:

$$\sum_j T_{ij} = O_i \qquad\qquad (11.4)$$

$$\sum_i T_{ij} = D_j \qquad\qquad (11.5)$$

Following a logical structure in the numbering of the zones, the modeller can proceed to analyse the common structures in areas of the O-D matrix (Figures 11.4 and 11.5):

- Trips with origin and destination in the same zone or intrazonal trips correspond to the diagonal of the O-D matrix.
- *Trips between zones in the study area*: These usually represent the biggest number of elements in the matrix and are distributed on both sides of the diagonal's upper part.
- Trips between zones within the study area and zones outside it are usually grouped in the lower left and upper right extremes.
- Trips between external zones that may or may not pass along part of the study area network are grouped on both sides of the diagonal at the lower part of the matrix.

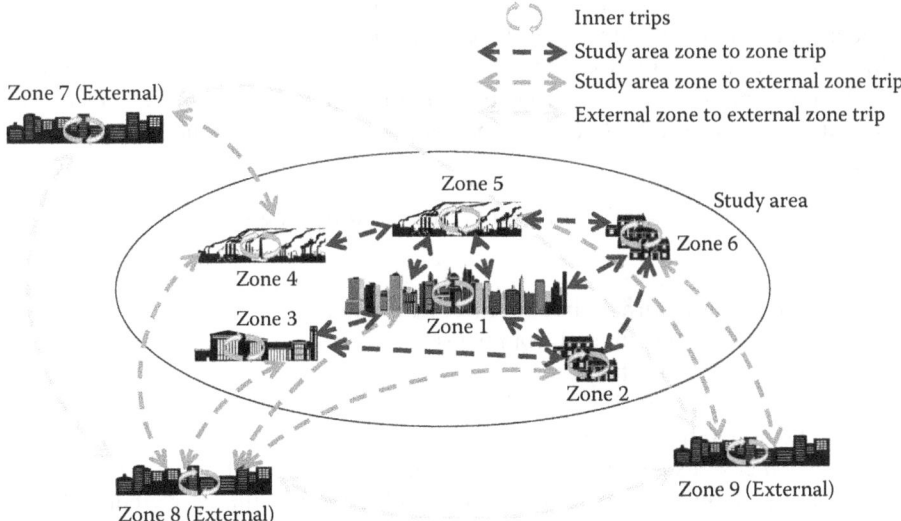

FIGURE 11.4 Types of trips between different zones.

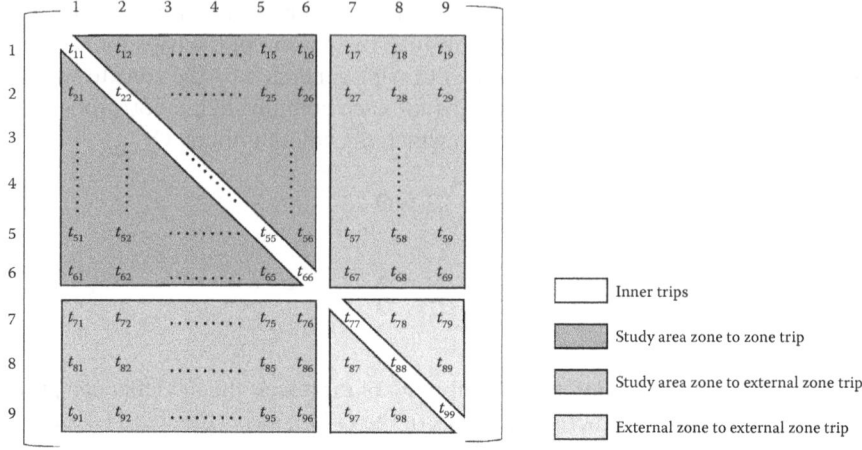

FIGURE 11.5 Types of trips in an origin–destination matrix.

Internal trips are not normally considered in modelling and it is a common practice for simulation software not to assign them to the network if the movement stays within the same zone. Nevertheless, trip distribution models are ideal for modelling journeys made between zones in a study area (also known as internal zones). This leaves the trips where the origin and/or destination are external zones: a common practice in these cases consists of considering these trips as being external to the modelling process, for example, asking interviews along a street at specific entrance and exit points along the boundary round the study area (de Dios Ortúzar and Willumsen 2011). The resulting matrix between the external and external–internal trips is estimated using growth factor methods and, more specifically, the Furness method.

Finally, one of the influential factors on the spatial distribution of journeys, or the variable number of trips between O-D pairs, is the perceived cost between the zones. This cost can be measured in time, money or distance in such a way that its value will be the sum of a series of variables trying to represent the disutility of making a specific journey, usually known as the generalised cost, the generic expression of which is:

$$c_{ij} = a_1 t_{ij}^v + a_2 t_{ij}^w + a_3 t_{ij}^t + a_4 t_{nij} + a_5 F_{ij} + a_6 \phi_j + \delta \qquad (11.6)$$

where:
t_{ij}^v is the on-board journey time to travel from i to j
t_{ij}^w is the walking time to and from the stop
t_{ij}^t is the waiting time at the stop
t_{nij} is the transfer time, if required
F_{ij} is the fare to travel from i to j
ϕ_j is a *end* cost (usually the cost of parking), associated with the trip from i to j
δ is a modal penalty, a parameter representing the remaining attributes that are not included in the generalised journey cost, for example, safety, comfort and so on

$a_1 \cdots a_6$ are the weights associated to each element of the cost; these weights have their appropriate corresponding dimension to convert each attribute into the same unit, for example, money or time

If the generalised cost is expressed in monetary units, then a_1 is sometimes interpreted as *the value of time* (or more precisely the *value of time* on board the vehicle), when the unit is money/time. In this case, a_2 and a_3 represent the value of time on foot and waiting time, respectively; in many practical cases, this is perceived as two or three times the value of a_1. If the generalised cost is expressed in monetary terms, then a_5 is normally fixed equal to one.

11.3.2   FROM PRODUCTION/ATTRACTION TO ORIGIN/DESTINATION

A very important aspect to consider is that the distribution models have been developed under the hypothesis that each trip had a generation and an attraction as its termini. Essentially, the models tie or relate the generations to the attractions. Generation is always the home in the case of HB trips; however, the origin of such journeys is only the home for those journeys taking place towards the work place (or the place of study, shopping, etc.) but in the return journey, home now becomes the destination.

In order to assign a trip matrix to the network, the production/attraction vector needs to be transformed into origins/destinations. In the case of daily matrices, a homogenous distribution of 50% between origins and destinations can be assumed where each generated/attracted journey takes place in each direction during the day. However, with O-D matrices covering short time periods, some journeys could possibly be made in the direction of generation towards attraction, whereas others are only made in the opposite direction. Three different approaches can be used to address this problem. The first consists of creating a matrix for only one purpose of journey, normally *work*, thereby assuming that these journeys are made only in one direction; this happens, for example, with the morning trips going to work from generation to attraction. The survey data should be used to correct such cases as shift work or trips made for other purposes during rush hour. A second approach consists of directly using the survey data to determine the proportion of each matrix for each purpose, which is considered as appropriate for the part of the day in question. For example, 70% of a typical morning rush hour matrix may represent generation to attraction trips and only 15% attraction to generation trips. The predominant purpose in each period of the day can be found from the household survey itself and a trip diary (Figure 11.3). If sufficient survey information is available, a commonly used third approach is to calibrate directly origin and attraction journeys by zone, using the same methodologies as those explained in Section 11.2.2. In this way, the final result of the generation stage would directly be the input or the constraints of the distribution model.

11.3.3   GRAVITY MODEL

The gravity model is one of the more frequently used methods for simulating transport. As explained in Chapter 5, it is derived from the analogy with Newton's law of gravity.

The model establishes that journeys made between a pair of zones are proportional to the production/attraction capacity of the zones and inversely proportional to the generalised cost. The classic functional form of the gravity model takes an exponential cost function (Wilson 1970), resulting in:

$$T_{ij} = A_i O_i B_j D_j \exp\left(-\beta c_{ij}\right) \tag{11.7}$$

where:
A_i and B_j are balancing factors for each zone referring to origins and destinations, respectively
O_i are journeys with origin in zone i
D_j journeys with destination in zones j
c_{ij} is the generalised cost between the pair of zones $i - j$
β is a parameter to be calibrated

Other functional forms that have been used as $f(c_{ij})$ are:

$$f(c_{ij}) = c_{ij}^{-n} \quad \text{Potential function}$$

$$f(c_{ij}) = c_{ij}^{n} \exp(-\beta c_{ij}) \quad \text{Combined function}$$

All these specifications of the gravity model can be derived using the well-known maximum entropy methodology, see Wilson (1970); in other words, given a series of origins and destinations, the gravity model is shown to determine the most probable trip matrix that minimises the total cost of travelling on a transport system. Various authors have since used the approach of maximum entropy to derive more sophisticated gravity distribution models to solve multiobjective optimisation problems where the fit with the observed data varies as a function of the level of spatial aggregation in the chosen zoning system (de Grange et al. 2010).

Obviously, the value of the calibration parameters n and β of the previous functional forms influences the final trip matrix solution. The values of $f(c_{ij})$ are shown in Figure 11.6 for different scaling parameter values. The resulting curves can be seen to be totally different to each other, even for the same functional form, which needs to be carefully considered when calibrating the model.

These parameters are calibrated assuming the model reproduces as accurately as possible the so called trip length distribution (TLD) observed in the sampling process. These distributions are variables that depend on the mode of transport being considered and, logically, on the definition of the cost or journey distance intervals, following the modeller's criteria.

11.3.4 TRI-PROPORTIONAL MODEL

In a more general version of the gravity model aimed at achieving a better fit to the TLD, the journey costs can be aggregated into a small number of cost intervals (10 or 15), indicated with the superscript m making the friction function to take the following form:

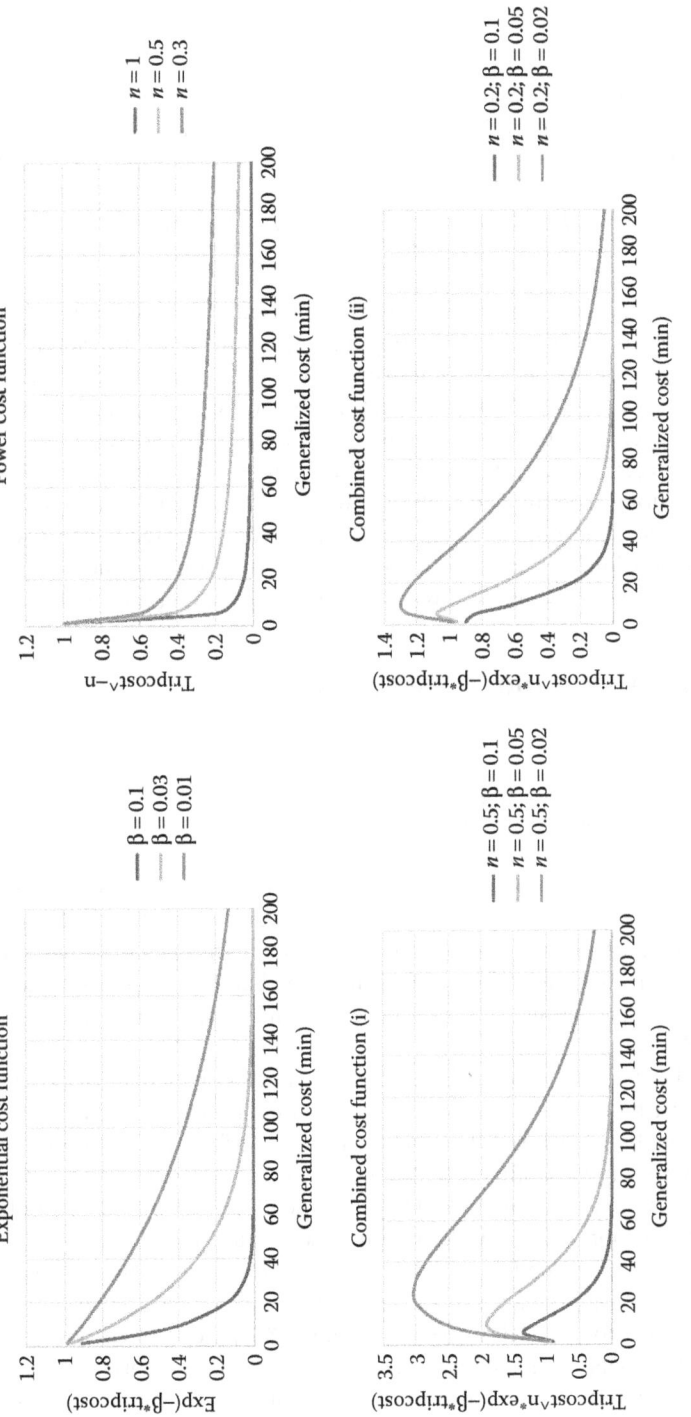

FIGURE 11.6 Different cost functions with different calibration parameters.

$$f(c_{ij}) = \sum_{m} F^{m} \delta_{ij}^{m} \tag{11.8}$$

where F^{m} represents the average value of cost for interval m, while δ_{ij}^{m} is equal to 1 if the cost of travelling between i and j is within interval m and equal to 0 in other cases.

The following expression can be reached by, once again, using maximum entropy:

$$T_{ij} = A_{i}O_{i}B_{j}D_{j} \sum_{m} F^{m} \delta_{ij}^{m} = a_{i}b_{j} \sum_{m} F^{m} \delta_{ij}^{m} \tag{11.9}$$

representing the gravity model with a function based on cost intervals. The calibration of a model like this requires finding adequate values for the factor F^m for each cost interval so that the number of journeys for a given distance will be as close as possible to the observed number. In fact, this procedure is very similar to the matrix expansion problem that has to satisfy the generated and attracted totals.

The process can be started by taking a unit value for the cost factors and then correcting them along with the parameters a_i and b_j, until the trip totals satisfy the TLD constraints. Thus, the bi-proportional algorithm can be extended to also consider this third dimension (cost interval) and use a tri-proportional method to calibrate the model with successive iterations balancing the rows, columns and cost intervals.

11.4   MODAL DISTRIBUTION MODELS

11.4.1   Basic Concepts

The main goal during this stage is to determine, given a group of trips between an origin and a destination, the proportion of journeys made using each mode based on a range of their characteristics and user preferences.

The factors influencing modal choice can be classified into three groups: characteristics of the people making the journey (car availability, driving licence, family size, personal constraints or restrictions on activity planning, etc.); characteristics of the journey (purpose, time of day, etc.); and the characteristics associated with the mode of transport (waiting, journey and access times, fare, comfort, regularity, suitable time table, etc.).

Empirical observations have shown certain habits in user behaviour when choosing a transport mode, for example, the trips made late in the day are less probably made using public transport, whereas journeys made for study purpose are strongly associated with car availability. The Figure 11.7 provides evidence from a household survey asked in Santander in 2009. They specifically represent the percentage usage of modes of transport according to trip purpose (Figure 11.7).

Modal distribution models can be classified into aggregated: those based on zonal data; and disaggregated: those generally based on individual data. Although the former were widely used in the 1970s and 1980s, they are currently being used as simplified models where the absence of data hinders the correct calibration of a disaggregated model. With respect to the latter, on a conventional level of transport modelling in a LUTI model, the more commonly used specifications are similar to

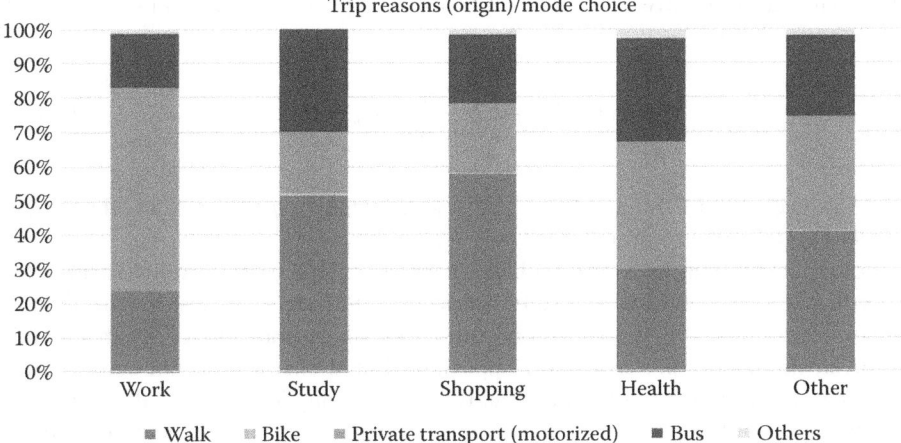

FIGURE 11.7 Modal choice according to trip purpose.

those that have already been described in Chapter 6 for the multinomial and nested logit models.

11.4.2 MULTINOMIAL LOGIT MODELS

From the foundations explained previously in Chapter 6, the choice probabilities of mode j and individual n are estimated using the following expression:

$$P_{jn} = \frac{\exp(\lambda_j V_{jn})}{\sum_{j=1}^{J} \exp(\lambda_j V_{jn})} \qquad (11.10)$$

Applied to a transport model, the probability is assumed to be related to the following expression:

$$P^k = \frac{\exp^{(-\beta \cdot (C_k + \delta_k))}}{\sum_m \exp^{(-\beta \cdot (C_m + \delta_m))}} \qquad (11.11)$$

where:

C_k is the cost of choosing mode k

δ_k is the modal penalty or other additional factors

The MNL model is estimated using the technique of maximum likelihood, the idea being to maximise the probability of obtaining the observed choices. The likelihood function can be expressed in the following way:

$$L(\theta) = \prod_{n=1}^{N} \prod_{A_j \in A(n)} (P_{jn})^{g_{jn}} \qquad (11.12)$$

The logarithm of the overall likelihood from a sample of independent observations can be expressed as:

$$\ln L = l(\theta) = \sum_{n=1}^{N} \sum_{A_j \in \underline{A}(n)} g_{jn} \ln P_{jn} \tag{11.13}$$

where g_{jn} is equal to 1 if individual 'n' chooses alternative 'j' and zero in other cases. Finally P_{jn} is the choice probability.

Note that different tests are available that are capable of finding the best model according to parameter-based results; a few examples are: the t test for the significance of parameter θ_k, the likelihood ratio test, the overall test of fit and so on. However, this model possesses two important limitations: it does not consider the presence of correlation between alternatives, which may lead to biased predictions (the blue bus-red bus paradox) and it does not take into account the heterogeneity of preferences among individuals.

11.4.3 Nested Logit Models

This model is a generalisation of the MNL model, which accepts the existence of correlated alternatives. If a simple choice situation between three alternatives, car, bus and metro, is considered, it can be represented with the following schematic (Figure 11.8).

Using Figure 11.8, each branch of the tree has a probability of reaching the corresponding alternatives. The probability of choosing one mode is found using the law of total probabilities, for example, $P(\text{Metro})=P(\text{Metro}|\text{TP})*P(\text{TP})$, where $P(\text{Metro}|\text{TP})$ is the probability of choosing the Metro given that the public transport nest has been chosen.

The solution proposes nesting the options available to a user, grouping together the alternatives that are in direct correlation on one level. Its structure is

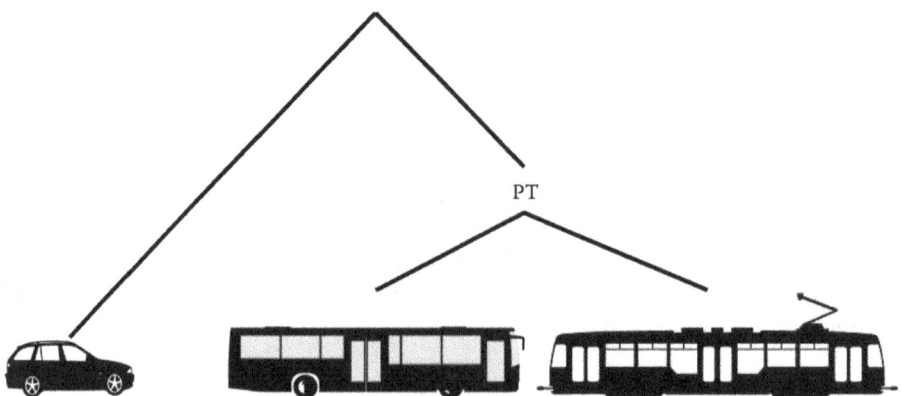

FIGURE 11.8 Possible structure of correlation between alternatives.

characterised by grouping sub-groups of correlated alternatives into nests or hierarchies. Each nest is represented by a compound utility possessing the following two components:

- One part that considers the expected maximum utility between the alternatives of the nest (EMU) as a variable, obeying the following expression:

$$\text{EMU} = \ln \sum_{A_j \in \underline{A}'(n)} \exp\left(\frac{\overline{V}_j}{\varphi}\right) \qquad (11.14)$$

where $\overline{V}_j$ is the utility of option A_j of the nest, except for the variables taking the same value for $\underline{A}'(n)$.

- Another component that considers the vector $\underline{W}$ of the attributes common to all the members of the nest.

Thus the compound utility of the nest is given by:

$$\tilde{U}_b = \varphi\text{EMU} + \underline{\alpha}\underline{W}$$

where φ and α are parameters to be estimated.

If the proposed hierarchical structure is correct, then $0 < \varphi \leq 1$. In the case where $\varphi = 1$, the hierarchical model collapses, signifying that for this particular case it is mathematically equivalent to MNL, which should therefore be preferred instead. If the model has more than one hierarchical level, the condition that needs to be fulfilled is $0 < \varphi_1 \leq \varphi_2 \leq \ldots \leq \varphi_n \leq 1$, in which φ_1 is the parameter of the EMU of the most internal nest and φ_b is the parameter of the upper nest.

11.5   ASSIGNMENT MODELS

11.5.1   INTRODUCTION

The process of loading a network with the demand for each mode of transport and representing the resulting transport flows across the same network is known as assignment. The main goal of this stage is to reproduce traffic, passenger and/or goods flows over the transport network. For example, the volume of traffic along each network link or the volume of passengers along each stretch of a public transport line can be found.

The assignment process involves a route choice model, given a group of available routes for making a journey. This model can either be of the deterministic or stochastic kind. Hypotheses can be assumed about how the network flow levels will affect journey times. If journey costs and times are supposed to be constant and independent of network flows, then the assignment model is *without congestion*. With the more common hypothesis, the user experiences longer journey times with increased network flows. In this case, the assignment model will be with congestion or equilibrium (Table 11.1).

TABLE 11.1

Classification of Assignment Models

	Deterministic	Stochastic
Constant costs	All or nothing	Stochastic assignment (dial, burrell…)
Costs dependent on flow (congestion)	User equilibrium (deterministic)	User equilibrium (stochastic)

Assignment is therefore the process of linking supply and demand, through which, given a modal demand represented by an origin/destination matrix and a transport system codified in a network model, the journeys are propagated across the network and result in a traffic flow pattern.

The most common flow-dependent cost hypothesis presents a reciprocal relationship between the flow pattern and the costs involved in moving between zones. This, in turn, has repercussions on accessibility, the trip generation phase and trip distribution and could modify the values of the utility functions of each mode during the modal distribution stage. This is why during situations of network congestion, these costs will vary significantly and the model may move into a cycle of iterative repetition between the different stages until a consistency of flows and costs is reached. This could mean that high levels of congestion, recurring to simultaneous modal split-assignment or distribution-modal split-assignment models, are the most commonly used.

11.5.2 Representation of the Road Network

The first step in obtaining a transport assignment model is the construction of the network. For the modelling purposes, the road network is represented by a graph $G(N, A)$, where N is the group of nodes and A is the group of links. So each network link $a \in A$ will have at the very least a series of attributes, such as:

- Name of the road or identification of the link
- Starting node and finishing node
- Length (l_a)
- Free flow speed ($v0_a$)
- Capacity (K_a)
- Link cost function

Figure 11.9 shows an example of network coding for a simple junction, disaggregating or not the characteristics of the movements at the nodes.

The public transport network, $\bar{G} = (\bar{N}, L)$, is formed of a group of nodes $\bar{N}$, subgroup of N and a group of public transport lines L, each of which is defined by a sequence of nodes where the passengers can board or alight the service.

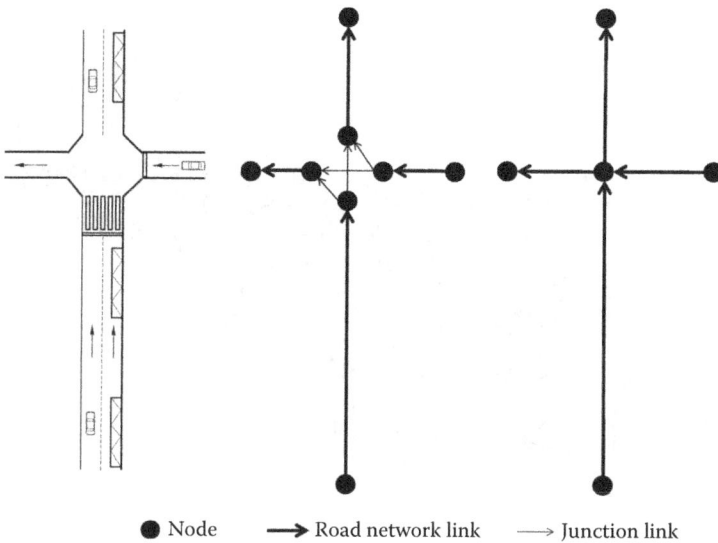

FIGURE 11.9   Examples of coding the road network.

Service networks, $G_{\bar{m}}(N_{\bar{m}}, S_{\bar{m}})$, are defined for each mode $\bar{m}$ of public transport, where $N_{\bar{m}}$ is the group of nodes and $S_{\bar{m}}$ is the group of *public transport links* (route sections) for mode $\bar{m}$. A route section is a portion of a route between two consecutive transfer nodes and has an associated group of lines that are equally *attractive* for the users (de Cea and Fernandez 1993).

Finally, the centroids of each zone need to be connected to the network using the so called link connectors, which completes the network model.

Figure 11.10 provides the urban network of the city of Santander (Spain) as an example to highlight the centroids, nodes and link connectors.

11.5.3 COST FUNCTIONS ON LINKS

As mentioned in the introduction to this section, it is assumed that the user of a transport mode bears a cost (or a time) associated with their journey along a network link that will be a function of the vehicle flow along the link, the capacity of the link and the speed limits on each road. Therefore, c_a is defined as the average journey cost on link a for users of private transport, as a function of vehicle flows (V_a), road capacity (K_a) and speed limits ($t0_a$) for link a. Multiple link cost functions can be found in the literature; one of the better known and widely accepted is the BPR function, expressed generically as:

$$c_a = t0_a \cdot \left(1 + \alpha \cdot \left(\frac{V_a}{K_a}\right)^{\beta}\right) \tag{11.15}$$

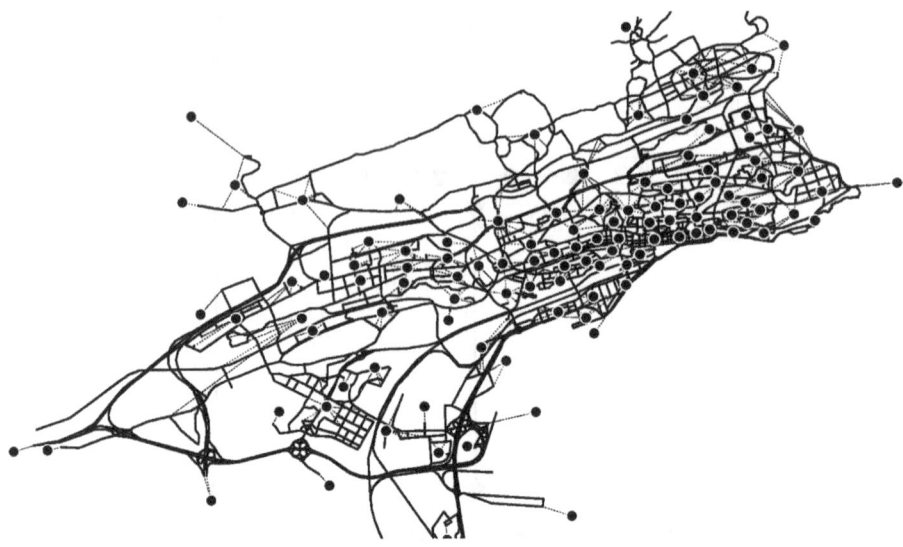

FIGURE 11.10 Coded transport network for Santander.

where y and β are parameters to be calibrated. The cost function has to be continuous and not decreasing. This expression can be seen to fulfil both conditions. Figure 11.11 provides a graph of a BPR cost function for an initial journey time of 15 seconds and a capacity of 1700 vehicles per hour for different values of the calibration parameters. The traffic counting analysis on certain streets and the use of floating vehicle techniques may be adequate for its calibration.

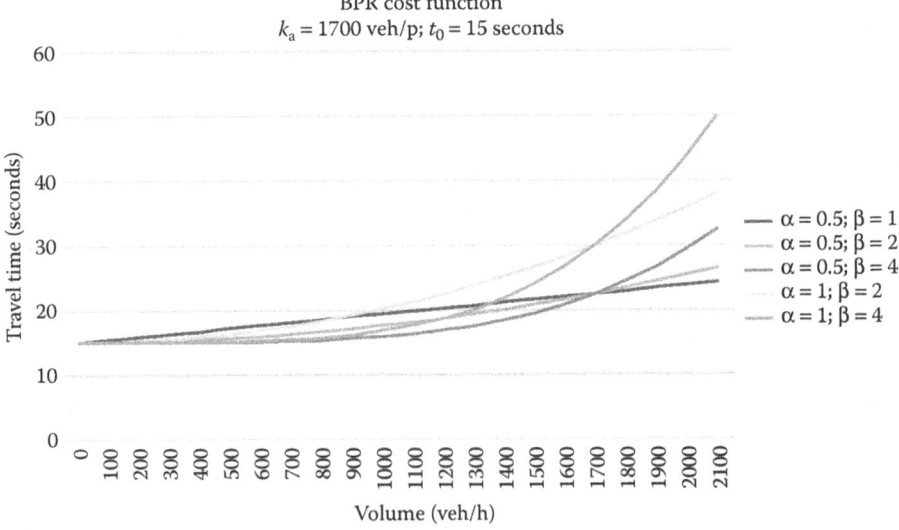

FIGURE 11.11 Example of the BPR function.

The cost functions of the public transport network differ slightly from those specified for private transport in which they implicitly consider a greater number of stages along the journey. The cost functions of public transport mainly cover the fare, the in-vehicle journey time and the waiting time. A typical approach to formulating these functions can be found in Chriqui and Robillard (1975).

11.5.4 Assignment of User Equilibrium

The classic assignment model within a transport model corresponds to a problem of traffic equilibrium across a network (roads) on which multiple categories of users interact.

The user equilibrium solution (UE) to the assignment problem is based on Wardrop's first principle (Wardrop 1952) which establishes that *for each category and for each origin-destination pair of a journey, no traveller can reduce their journey time (cost) by unilaterally changing their route on the network*. The routes that have been used for each O-D pair will have the same operating costs, and the routes that were not used will also have equal or greater operating costs.

The hypotheses adopted for UE are as follows:

- The user has complete knowledge about the network and the costs of the different routes.
- The journey time along a link is only a function of the flow along the link.
- The cost functions (or journey times) are continuous and increasing.

The constraints on the problem are those known as conservation and continuity of flows, as well as that of no negativity. In other words, the demand for a particular O-D pair will be distributed across the different routes connecting this pair whose cost is minimum and equal to that of equilibrium; the flow along a link is the sum of the flows along routes that cross this link and the traffic flows are equal to and greater than 0, as shown in the following:

$$g_w = \sum_{p \in P_w} h_p, \quad \forall w \in W \tag{11.16}$$

$$v_a = \sum_{p \in P} h_p \cdot \delta_{ap}, \quad \forall a \in A \tag{11.17}$$

$$h_p \geq 0, \quad \forall p \in P \tag{11.18}$$

The solution to the equilibrium problem can be put forward as the next equivalent optimisation problem (Beckmann et al. 1956):

$$\text{Min} : Z(f) = \sum_l \int_0^{f_l} c_l(x) dx \tag{11.19}$$

subjected to the constraints set out previously.

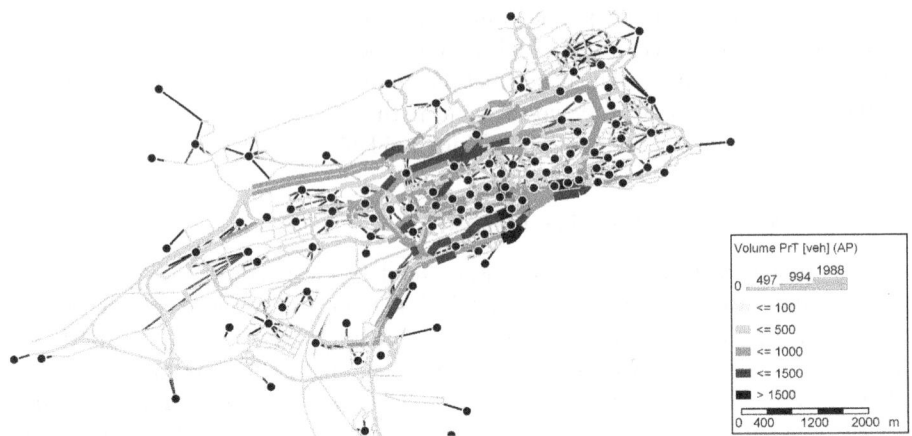

FIGURE 11.12   Example of network assignment.

In the case of public transport networks, where the hypothesis of not considering congestion may be more common, the assignment problem is similar to that presented but for the cost functions of the route sections. The model assumes that the users choose, from all the possible routes that join a determined pair of nodes on a public transport network, the route that minimises their total cost (fare + in-vehicle journey time + waiting time + access time). The flows of passengers along each section are proportionally assigned to the lines that form it according to their relative nominal frequency. If congestion is being considered, the problem becomes a problem with cost interaction, whose Jacobian is not diagonal and not symmetrical and which is outside the scope of this chapter.

One of the more frequently used algorithms for solving the traffic equilibrium model in the private transport assignment model is that of Frank–Wolfe (Frank and Wolfe 1956), which is a variation of the well-known gradient method.

Figure 11.12 shows an example of traffic assignment used in the Santander model. The darker shaded links represent those with a higher saturation index (intensity/capacity).

11.5.5   Stochastic User Equilibrium

The UE explained earlier, based on Wardrop's first principle, assumes that all the users perceive the cost in the same way (deterministic route choice). Sheffi (1985) formulated an equivalent problem in which, considering stochasticity in the route choice, the users perceive the costs of different routes between O-D pairs differently. Therefore, analogous to the utility of the probabilistic models in previous stages, the cost of route k can be expressed as:

$$U_k^i = V_k^i + \varepsilon_k^i = -\beta g_k^i + \varepsilon_k^i \tag{11.20}$$

The probability of choosing a certain route k can be determined as the probability that its cost (or journey time) will be perceived as lower than that of the alternative routes. This principle, that in the models without congestion gives place to the stochastic assignment process or SNL, is what introduces a cost function dependent on network flows into the equilibrium models. Stochastic user equilibrium (SUE) can now be described as *no user can reduce their perceived journey time by unilaterally changing route*. In fact, UE can be considered as a particular case of SUE in which the variance of the perceived journey time is 0.

The algorithm usually employed to solve the problem of SUE is known as the method of successive averages. This algorithm has an inconvenience that needs to be considered: its difficulty of convergence on very congested networks. What could happen on these kinds of networks is that the error component in the cost term becomes almost negligible compared to the high journey costs. In these cases, the solution of the SUE converges on the equivalent to the UE, meaning that the problem can be addressed as a deterministic problem, directly solving the UE by using the more efficient Frank–Wolfe algorithm.

REFERENCES

Beckmann, M. J., C. B. McGuire, and C. B. Winsten. 1956. *Studies in the Economics of Transportation*. New Haven, CT: Cowles Commission for Research in Economics, Yale University Press.

Chriqui, C., and P. Robillard. 1975. Common bus lines. *Transportation Science* 9 (2): 115–121. doi:10.1287/trsc.9.2.115.

de Cea, J., and E. Fernandez. 1993. Transit assignment for congested public transport systems: An equilibrium model. *Transportation Science* 27 (2): 133–147. doi:10.1287/trsc.27.2.133.

de Dios Ortúzar, J., and L. G. Willumsen. 2011. *Modelling Transport*. Hoboken, NJ: John Wiley & Sons.

de Grange, L., E. Fernández, and J. de Cea. 2010. A consolidated model of trip distribution. *Transportation Research Part E: Logistics and Transportation Review* 46 (1): 61–75.

Frank, M., and P. Wolfe. 1956. An algorithm for quadratic programming. *Naval Research Logistics Quarterly* 3 (1–2): 95–110.

Sheffi, Y. 1985. *Urban Transportation Networks*. Englewood Cliffs, NJ: Prentice Hall.

Wardrop, J. G. 1952. Some theoretical aspects of road traffic research. *Proceedings of the Institute of Civil Engineers* , London, UK.

Wilson, A. G. 1970. *Entropy in Urban and Regional Modelling, Monographs in Spatial and Environmental Systems Analysis 1*. London, UK: Pion.

Section IV

Land Use–Transport Interaction Models Considering Spatial Dependence

In this section, we will review the advantages of considering the effects of the possible presence of spatial dependence in the data used to feed land use–transport interaction (LUTI) models. Chapter 13 will also describe an example of calibration of a combined LUTI model without considering and considering the presence of spatial dependence between the observations. Finally, an example of using this type of model for the simulation of urban and transport planning policies will be provided.

12 Spatial Dependence in LUTI Models

Rubén Cordera and Luigi dell'Olio

CONTENTS

The land use–transport interaction (LUTI) models described in Chapters 1 through 11 and especially the techniques that are more commonly used such as multilinear regression, multinomial logit (MNL) discrete choice models and gravity models start from the hypothesis of independence between observations or choice alternatives (Gujarati and Porter 2009, Hensher et al. 2015). This starting hypothesis assumes that the random residuals forming part of the stochastic component of the models are not correlated either directly or by the effect of autocorrelation between different observations of the variable to be estimated (e.g. trips generated or real estate prices). If this hypothesis is not fulfilled in the data, then the parameters of the model could be biased or inefficient (Ward and Gleditsch 2008). If the parameters are biased, they will not be representative of the population and the predictions made by the model will not be accurate. If the parameters contain high standard errors, then the predictions made by the model will be imprecise. Both problems are relevant although the problem of inaccuracy is more serious as it could lead to the drawing of incorrect conclusions from the simulations made using the model. It is advisable for the modellers to be aware of the possible presence of this problem and to avoid it by running the pertinent tests or modifications when specifying the functions.

The following sections will examine in greater detail the phenomenon of spatial dependence in cross-sectional data or panel data and will explore their causes and the effects they have on the estimation of econometric models. Sections 12.2 through 12.4 will address, respectively, the problem of spatial dependence in location models, hedonic regression models and trip generation models providing three examples where the presence of these kinds of spatial effects could generate inaccurate or imprecise models. Finally, Section 12.5 will present the conclusions about how to address this problem in LUTI models and describe the improvements that the use of spatial econometric techniques can bring to their estimation.

12.1   SPATIAL DEPENDENCE IN CROSS-SECTIONAL DATA

LUTI models require large amounts of data for their calibration. Among all this information, cross-sectional data that is zonal or point data for a determined moment in time is the more frequently used. Similar to time series data, cross-sectional data can show situations of spatial dependency between observations, which violate the traditional hypotheses behind regression and discrete choice models based on the idea of independence between observations or choice alternatives (Anselin and Rey 2014). The fundamental difference between the dependency of time series data and cross-sectional data with a spatial dimension is that when the former shows a clear direction of dependence, the spatial data can present dependency with multiple neighbouring observations and possible feedback effects. The essence of the spatial models is therefore one of the explicitly considered spatial component of the observations whilst considering the so-called Tobler first law of geography (Tobler, 1970): *everything is related to everything else, but near things are more related than distant things*. This is formally presented as:

$$\text{Cor}(y_i, y_j) = E(y_i, y_j) - E(y_i)E(y_j) \neq 0 \tag{12.1}$$

where y_i and y_j are observations of the same variable in different positions in space.

Spatial econometric models started to appear in the 1970s in an attempt to address data on a regional scale (Anselin 2001). This type of model was later applied to the growing number of problems related with, among other things, urban issues, the housing sector (Armstrong and Rodríguez 2006, Long et al. 2007, Ibeas et al. 2012) and economic geography (Fingleton 2001, 2003). In all these fields, the use of data with a clearly spatial component is an essential part of the modelling process, and it quickly underlines the importance of addressing it by specifically considering its peculiarities, as had occurred with time series data in other research fields. Added to this, the greater availability of georeferenced information has also helped in the development of LUTI models (for more details, see Chapter 1).

The spatial effects present in cross-sectional data or in panel data can be classified into two general types:

- *Spatial dependence*: The existence of a relationship between neighbouring observations in spatial terms
- *Spatial heterogeneity*: The structural instability in a study area either in variances of the model errors or in the estimated parameters

This chapter will not address the subject of spatial heterogeneity, the treatment of which requires the use of other specific econometric techniques. By contrast, the situation of spatial dependence between observations can be addressed using diverse methods according to the type of model being used.

Where linear regression is used, spatial dependence has normally been addressed using two methods. First, by specifying a spatial autocorrelation model, in other words, where the dependent variable of neighbouring spatial observations becomes the explanatory variable of the local variable. Mathematically, the regression model is therefore specified as

$$y = \rho W y + X\beta + \varepsilon \qquad (12.2)$$

where:

y is a vector containing the data of the dependent variable

β is a vector of parameters to be estimated

X is a matrix containing the information of the independent variables

ε is a vector of independent and identically distributed errors (IID)

As can be seen from the equation, the dependent variable is also specified on the right side of the equal sign where ρ is a parameter of spatial autocorrelation and W is a matrix with the specification of which observations are considered neighbours of each local observation. The elements of this matrix are normally standardised by rows so that in each row $i \sum_j w_{ij} = 1$. In this way, the autocorrelation with the neighbouring dependent variables can be interpreted as a weighted mean (Anselin 1988b). In the literature, this model is known as a spatial autoregressive model (SAR). The presence of the dependent variable on both sides of the equation generates the direct modelling of the phenomenon of feedback derived from the spatial dependence between observations. This means that the SAR model can no longer be estimated using ordinary least squares because of the endogeneity generated by vector y and, therefore, alternative estimation methods such as maximum likelihood need to be used.

Second method for addressing spatial dependence between neighbouring spatial observations is to model it from the spatial correlation of the regression residuals. In the literature, this method has been called the spatial error model (SEM) and is only appropriate when attempting to correct the possible bias that spatial dependence may introduce into the data. However, on its own it does not model the process of spatial dependence as the specification of the SAR model does. Nevertheless, the parameters of the SEM model can be estimated using ordinary least squares without showing bias. The SEM model is specified as

$$y = X\beta + u \qquad (12.3)$$

$$u = \lambda W u + \varepsilon \qquad (12.4)$$

where:

λ is a parameter of autocorrelation of the errors μ

ε is a vector of IID errors

In this model, each location is as much a function of the independent variables as of the errors of the neighbouring locations.

It is also possible to combine the SAR and SEM models in order to generate a model that is capable of capturing the spatial autocorrelation in both the dependent variable and in the error term.

$$y = \rho W_1 y + X\beta + u \qquad (12.5)$$

$$u = \lambda W_2 u + \varepsilon \qquad\qquad (12.6)$$

The model, known as SAC in the literature, can therefore be specified with two different neighbourhood matrices for the dependent variable (W_1) and the error term (W_2) (LeSage and Pace 2009). This kind of model can be useful when the residuals of the model are found to be correlated even when a SAR model has been applied.

The W matrix, used to specify the neighbouring relationship between observations, can be defined using various criteria depending on whether the data are zonal or point in nature. There are four methods most commonly found in the literature for defining the matrices (Figure 12.1): queen, tower, a predetermined number of nearest neighbours and maximum distance. In the queen contiguity case, all the observations sharing an edge or a vertex with the local observation are considered to be neighbours. In the tower contiguity case, only the locations sharing edges with the reference observation are considered to be neighbours. Both methods are, therefore, appropriate for determining proximity in cases where the data have a polygonal form of spatial expression. The maximum distance and K nearest neighbours methods can also be applied to point data. In the former, a maximum distance is applied from the local observation to any possible neighbours, whereas in the latter a predetermined number of nearest neighbours is specified. The queen-type and tower-type neighbourhood matrix can only be applied to point data if they are previously converted using techniques such as Thiessen polygons (Brassel and Reif 1979). In all these cases, proximity can be of an order above one, in other words, observations can be considered as neighbours of the observations which are in turn neighbours of the local observation, up to the desired order.

Extensive debate can be found in the specialised literature about the impact that the specification of the chosen proximity matrix has on the parameters estimated

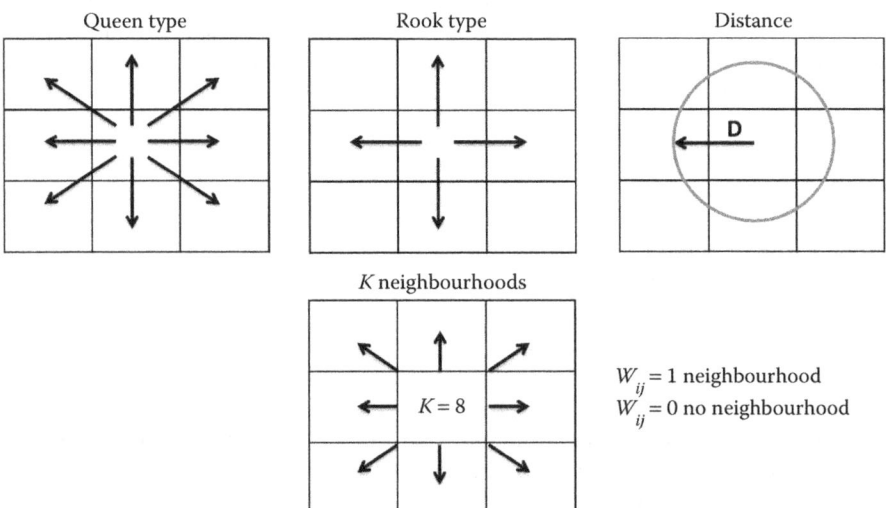

FIGURE 12.1 Methods for determining proximity between observations.

in the model. LeSage and Pace (2010a) used different indicators to measure the correlation between proximity matrices. These authors showed how the specification of W has a small influence on the estimated parameters if the model is well specified and its parameters are interpreted correctly from the true partial derivatives. Only the matrices that differ considerably in terms of the average neighbours found for each observation, for example, a specified distance matrix with a very wide radius versus a queen-type matrix, could imply clearly different estimated parameters.

Situations of spatial dependence between observations can also occur with discrete choice models. Discrete choice models (see Chapter 6) and specifically MNL models are founded on the hypothesis of identical variance and independence between the stochastic errors of the different alternatives, in other words, the variance/covariance matrix associated with a MNL model is of the following kind:

$$\sum = \sigma^2 \begin{bmatrix} 1 & \cdots & 0 \\ \vdots & \ddots & \vdots \\ 0 & \cdots & 1 \end{bmatrix} \tag{12.7}$$

This hypothesis can be restrictive in many choice situations. In the field of spatial choice, spatially close zones can share attributes that are not measured by the systematic utility of the alternatives, which can carry a certain degree of spatial dependence in the random component of the utility. This correlation is addressed by estimating more complex models than MNL, models which allow some types of correlation between alternatives to be specified. The more well-known models able to consider this correlation are nested logit (NL), cross nested logit (CNL) and the error components model (EC), all of which have been described in Chapter 6.

Before applying techniques with the ability to consider the presence of spatial dependence between observations or alternatives, the researcher must first check whether they are present and if so, to what degree. Specific indicators can be used for this purpose. Spatial correlation indicators can be split into two large categories: (1) global indicators that examine the presence of correlation for all the sample and (2) local indicators that attempt to find the specific places where this correlation appears in the study area. The most well known of the global indicators is the Moran I test (Moran 1948). This test is very similar to the Durbin–Watson statistic applied to time series analysis but adapted to a two-dimensional situation. The formula for the Moran I indicator is as follows:

$$I = \frac{n}{\sum\limits_{i=1}^{i=n}\sum\limits_{j=1}^{j=n} W_{ij}} \cdot \frac{\sum\limits_{i=1}^{i=n}\sum\limits_{j=1}^{j=n} W_{ij}(x_i - \overline{x})(x_j - \overline{x})}{\sum\limits_{i=1}^{i=n}(x_i - \overline{x})^2} \tag{12.8}$$

where W_{ij} is a neighbourhood matrix between observations, which takes a value of 1 if the observations are considered to be neighbours and 0 if not. The variable for examining the presence of spatial correlation is x, and n is the total number of observations in the sample. The W_{ij} matrix is typically standardised to facilitate the calculations by making $W_{ij} = w_{ij} / \sum w_{ij}$.

In order to check the presence of a significant degree of autocorrelation in the data and, therefore, the suitability of applying a conventional linear regression model, the variable x to be examined using the Moran index will be the residual from a conventional linear regression.

Whether the resulting Moran I is significantly different to a critical value or not can be calculated using its z value by

$$z_I = \frac{I - E[I]}{\sqrt{V[I]}} \tag{12.9}$$

where $E[I] = -1/(n-1)$ and $V[I] = E[I^2] - E[I]^2$. The null hypothesis of the test is the non-existence of spatial autocorrelation, the presence of which is determined by passing a certain critical value of Z_I for a determined confidence level. The Z_I statistic is derived analytically, leading to the assumption that the test distributes asymptotically normal. Therefore, the sample being used needs to be large, which could be problematic in studies that do not have a particularly detailed zoning or a large number of observations. Nevertheless, different simulations have shown that Z_I can have a distribution close to normal even with small sample sizes (Anselin and Rey 1991).

The Moran I indicator takes values between −1 and 1. An indicator value equal to 0 means a complete absence of any kind of spatial autocorrelation, a value of 1 indicates perfect positive spatial autocorrelation and a value of −1 perfect negative spatial autocorrelation.

Consider an example with three zones, as shown in Figure 12.2. The value shown inside the zones corresponds to their identifier, whereas the data about the variable of interest are found in brackets. The corresponding proximity matrix standardised by rows is as shown in Table 12.1 where the cells can only be neighbours of others if they share a common edge to east or west. Equation 12.8 can now be applied using this data, giving a Moran I of −0.48; in other words, negative spatial correlation is present derived from the proximity of the clearly different values for the variable of interest.

Both spatial regression models and discrete choice models considering correlation between the errors of the alternatives are normally estimated using maximum likelihood. Specialised spatial regression software can be used for this; examples are Geoda (Anselin and Rey 2014), the spdep package for the R programming language (Bivand et al. 2015) and, in the field of discrete choice models, software like

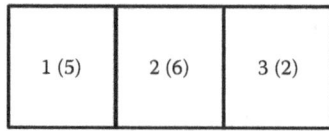

FIGURE 12.2   Example of spatial data in three zones.

TABLE 12.1
Example of a Neighbourhood Matrix
Standardised by Rows

W_{ij}	1	2	3
1	0	1	0
2	0.5	0	0.5
3	0	1	0

NLOGIT (Greene 2012) or Biogeme (Bierlaire 2003). Spatial regression models can also be estimated using alternative methods such as spatial two-stage least squares or the method of moments estimator.

If the spatial regression models have been estimated using maximum likelihood, then tests such as the likelihood ratio test (LR test) can be applied to check whether considering spatial autocorrelation in the dependent variable or in the residuals of the model significantly improves the fit compared to a non-spatial model (Ben-Akiva and Lerman 1985). Another series of tests such as Lagrange multipliers (LM) does not require the estimation of an additional spatial model and is also able to distinguish whether or not the estimation would be significantly improved by a SAR (LM_{lag}) or by an SEM model (LM_{err}) (Anselin 1988a). The LM test takes the following form (Anselin 2001):

$$LM_{lag} = \frac{\left[e'Wy/(e'e/n)\right]^2}{D} \qquad (12.10)$$

$$LM_{err} = \frac{[e'We/(e'e/n)]^2}{[tr(W^2 + W'W)]} \qquad (12.11)$$

where:
$D = [(WX\beta)'(I - X(X'X)^{-1}X')(WX\beta/\sigma^2] + tr(W^2 + W'W)$ is the vector of the residuals from the non-spatial regression model
W is the neighbourhood matrix
n is the number of observations

Both tests have a χ^2 distribution with a degree of freedom. The fit of the SAR or SEM spatial model will, therefore, be significantly better when passing a test value of 3.84 for a 95% confidence level.

12.2 SPATIAL DEPENDENCE IN POPULATION AND ACTIVITIES LOCATION MODELS

The population and activities location models presented in greater detail in Chapter 9 have normally been based on techniques such as gravity models, linear regression or, more recently, on MNL-type discrete choice models for the simulation of location choice made by agents involved in urban space.

With residential or economic activities location models, there may be complex substitution patterns between choice alternatives, and the hypothesis of uncorrelated errors assumed by the MNL model may prove too restricting. For example, let us imagine a household with a choice group of three available zones to locate into, two of which share common traits or are simply spatially closer together. Initially, the choice probabilities provided by the model could be determined as 0.70 for zone 1, 0.29 for zone 2 and 0.01 for zone 3. An improvement in the characteristics of zone 3 is going to be simulated, by which its choice probability increases to 0.1. In the MNL model, the probability of the other two zones has to be reduced by the same percentage, resulting in probabilities of 0.64 and 0.26, respectively. This implies that the improvement in the choice probability of the latter zone is a result of a 0.06 reduction in the choice probability of zone 1 and of 0.03 for zone 2, meaning the choice probability of zone 1 falls double that of zone 2. This may not be realistic in choice situations where, for example, zone 2 is more similar to zone 3 with characteristics that have not been accurately measured using the systematic utility. Where this occurs, there is a correlation between the errors of the alternatives, and it would be expected that the increase in the choice probability of zone 3 would correspond to a greater fall in the choice probability of zone 2 than in zone 1. Therefore, the pattern of substitutions between alternatives in the MNL model could show little realism in situations of correlation between alternatives such as those presented earlier.

This type of phenomena can be quite common in the context of spatial location because zones that are near to each other will almost definitely share common traits, as described earlier in Tobler's first law of geography. The literature provides examples of attempts to address this problem by using discrete choice models considering correlation in the residuals of the alternatives.

As mentioned previously, among the models available for considering correlation between alternatives are the NL, the CNL and the EC models. The NL model is the most inflexible of the three, as it requires the modeller to establish *a priori* a correlation structure between the alternatives where each alternative can only be correlated with one group of alternatives at the same time. The CNL model provides somewhat greater flexibility as it allows an alternative to be correlated at the same time with more than one group of alternatives. However, this requires the estimation of additional parameters of inclusion in the model. Finally, the EC model is able to specify flexible correlation structures by introducing shared ECs in the utility functions of different alternatives. Furthermore, the same alternative can have various ECs that allow it to be correlated with more than one group of alternatives.

Example 12.1: Residential Choice Models Considering Spatial Dependence between Choice Zones

To provide an example of using locational models considering spatial correlation between alternatives, in this case the study area will be the urban nucleus of Santander divided into 26 zones rather than the 42 zones of the metropolitan area proposed in Chapter 8 because of limitations with data availability.

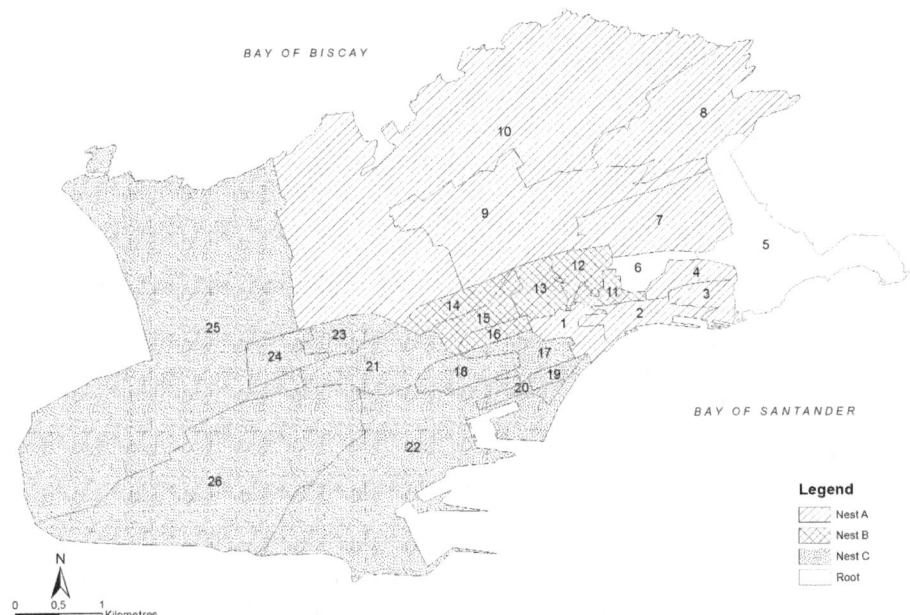

FIGURE 12.3 Proposed macro zones for considering correlation between alternatives.

A NL and a CNL model will be used in this example and they will be specified with a possible structure of correlation between the alternatives. A series of macro zones have been defined within which correlation between alternatives may exist derived from their closeness and the sharing of certain common characteristics that were not considered in the systematic utility function (Figure 12.3). In the case of the CNL model, the zones located on the edges of the macro zones were considered to belong to any of the nests. So, for example, zone 10 belongs to nest A, but it could also form part of nest B and nest C as it shares an edge with them.

The database used to estimate the models shows all the alternatives available for 534 households/heads of households. The NL and CNL models cannot be estimated directly if a database based on a random sample of alternatives is being used (as shown in Chapter 9) and is required to carry out certain modifications in the utility functions (Lee and Waddell 2010, Guevara and Ben-Akiva 2013).

The presence of correlation between alternatives and, therefore, the need to estimate a NL or CNL model can be shown by the parameter of each of the nests where the hypothetically correlated alternatives are grouped. If the parameter of each nest is not significantly different from 1, in the case of having performed the normalisation from the upper nest, then the nest can be removed as this means no specific covariance is generating correlation between the alternatives.

Table 12.2 shows the parameters estimated using maximum likelihood for a MNL model without considering correlation between alternatives, for a NL model and for a CNL model. All of them were estimated using the Biogeme software. Their *p*-value is shown in brackets below the estimated parameters to demonstrate

if they are significantly different from zero. The following variables were introduced into the model:

- *ACCA*: Gravity indicator of active accessibility to employment from each
  of the zones.
- *CT*: Journey time in minutes from the chosen zone to reach the employ-
  ment zone.
- *FORE*: Population with origins outside the European Union present in
  the zone.
- *IN*: Dummy variable with a value of 1 if the residential and employment
  zones coincide.
- *LNVIV*: Natural logarithm of the number of dwellings present in the zone.
- *PG*: Dummy variable with a value of 1 if the zone has special prestige.
  This variable has to be determined by the modeller.
- *PRI*: Average house price in euros of property in the zone.
- *SCH*: Number of schools in the zone.
- *WT*: Average waiting time for public transport in the zone.

The variables are also introduced as having an interaction with a dummy variable
with a value equal to one if the household has an income of over 2500 euros to
test for any differential effect in the parameters (e.g. ACCAH, CTH, etc.).

Table 12.2 shows how journey time to work was a clearly significant variable,
without having any different effect on the heads of high-income households.
Other variables that were clearly relevant in increasing the utility of locating in a
zone were the number of dwellings, the prestige of the zone (with opposite signs
according to income) and the price, without any differences found for household
with incomes of over 2500 euros.

In the case of grouping into nests in the NL model, only nest B showed the
presence of significant correlation between zones, although the parameters of
nests A and C were close to being significantly different from 1. While thanks to
the greater flexibility of the CNL model, all its nests were clearly and significantly
different from 1.

TABLE 12.2
Residential Choice Models: MNL, NL and CNL

Variables	MNL Model	NL Model	CNL Model
ACCA	.005	.005	.006
	(.56)	(.51)	(.33)
ACCAH	−.016	−.013	−.003
	(.38)	(.39)	(.79)
CT	−.102	−.089	−.068
	(.01)	(.01)	(.02)
CTH	−.000	−.000	−.000
	(.92)	(.99)	(.99)
FORE	−.000	−.000	−.000
	(.04)	(.04)	(.13)

(Continued)

TABLE 12.2 (*Continued*)
Residential Choice Models: MNL, NL and CNL

Variables	MNL Model	NL Model	CNL Model
FOREH	−.001	−.001	−.000
	(.12)	(.12)	(.07)
IN	.235	.214	.203
	(.29)	(.26)	(.14)
LNVIV	1.39	1.28	.531
	(.00)	(.00)	(.07)
LNVIVH	1.49	1.15	.685
	(.07)	(.08)	(.13)
PG	−.897	−.681	−.339
	(.00)	(.01)	(.12)
PGH	1.90	1.76	1.19
	(.00)	(.00)	(.00)
PRI	−.000	−.000	−.000
	(.04)	(.01)	(.00)
PRIH	−.000	−.000	−.000
	(.57)	(.50)	(.53)
SCH	−.087	−.053	.000
	(.05)	(.17)	(.98)
SCHH	.242	.215	.152
	(.00)	(.00)	(.00)
WT	−.132	−.097	−.069
	(.10)	(.18)	(.32)
WTH	−.013	.048	.061
	(.94)	(.76)	(.65)
NEST A	−	1.24	3.97
		(.11)	(.01)
NEST B	−	1.28	1.85
		(.05)	(.05)
NEST C	−	1.11	1.27
		(.28)	(.00)
ROOT NEST	−	1.00	1.00
Null log-likelihood	−1739.82	−1739.82	−1739.82
Log-likelihood	−1658.58	−1654.73	−1626.64
LR test null	162.48	170.17	226.35
LR test MNL/NL	−	7.69	56.18
N	534	534	534

While comparing the models' fit with the data (see LR test in Table 12.2), both the NL and CNL had a significantly better fit than the MNL model without considering the possible presence of spatial correlation between the different zones and with the same variables in the utility function.

12.3   SPATIAL DEPENDENCE IN HEDONIC REGRESSION MODELS

As seen in Chapter 10, LUTI models have sometimes included a sub-model capable of simulating the impact of the transport subsystem on real estate values (e.g. the presence of a new bus stop close to certain housing). The most commonly used method has stemmed from the estimation of hedonic models through simple linear regression. These models have been specified with property prices, or a transformation of them, as the dependent variable along with a series of structural, environmental and transport characteristics as independent variables. A typical functional form of these models is:

$$\ln(P) = X\beta + \varepsilon \qquad (12.12)$$

where:

P is the vector with the property selling price

X represents a series of independent variables corresponding to the different characteristics of the heterogenous product

β is a vector of the parameters to be estimated

ε is a vector of independent identically distributed errors between observations

However, various authors have detected that hedonic models that are generally estimated using cross-sectional data usually have spatially correlated residuals according to tests such as Moran I (Armstrong and Rodríguez 2006, Ibeas et al. 2012). This has led to the need to estimate hedonic models that capture this correlation in order to fulfil the hypothesis of independence in the residuals assumed in regression models.

On a substantive level, the presence of autocorrelation in conventional hedonic models can be derived from the reciprocal influence on nearby property prices. This is, therefore a diffusion effect where property owners see prices updating as a function of the prices being offered or realised in nearby properties. It may also be the case where a model has a specification problem that has resulted in the omission of a variable with a differential spatial effect. This situation can sometimes generate spatially correlated residuals, a problem that can usually be removed by the correct specification of the model. Where this solution is not possible, an SEM model could at least address the spatial correlation in the residuals. Finally, another possible situation is where markets are differentiated in the data. In this case, it may be necessary to recur to models that can address spatial heterogeneity and consider the presence of different spatial structural relationships.

Example 12.2: Hedonic Regression Models Considering Spatial Dependence between Observations

The models estimated in Example 10.1 will form the basis of demonstrating the estimation of a hedonic regression model considering a situation of spatial dependence in the data. However, for consistency with the residential choice model estimated in Example 12.1, only the data relating to the Santander urban nucleus will be used. The TRANS variable was also modified to adapt it to a smaller study area. Instead of specifying 400 m as the closest distance between a property and a bus stop, 200 m was chosen as being more appropriate for an urban environment.

Given that hedonic regression models are based on the hypothesis of independence between the residuals of the different observations, an advisable first step would be to check for correlation in the residuals generated by the model. This means checking whether the differences between the prices estimated by the model for each property and the real price provided by the sample show any kind of spatial correlation between observations.

To address this, the Moran I index was applied to the residuals of models S-1 and S-2 from Example 10.1. In this case, the models were estimated only with the data from the properties located in the city that requires a new neighbourhood matrix to be defined between observations. In this example, a queen-type neighbourhood matrix has been generated from the Thiessen polygons. This neighbourhood matrix provides an average of 15 neighbours for each observation. In a more detailed modelling process, the models could be estimated with different neighbourhood matrices to test for significant variations in the estimations. A total of 846 observations were used, which is a sufficiently high number to check if the test is significantly different from zero.

For the residuals of both models (Table 12.3), the test was significantly different from zero (see the *p*-value in brackets), thereby showing the existence of a significant spatial correlation in the residuals. The results of the LM test can be seen in Table 12.4. In both cases, the tests were clearly significant with the LM_{lag} test providing the higher values.

These indicators would, therefore, appear to show the suitability of estimating SAR and SEM models (Table 12.5) to discover if there are variations in the estimated parameters or in the standard errors and therefore in the significance of the different parameters. It is also of interest to check if the spatial models improve the fit of the conventional hedonic models.

TABLE 12.3
Moran I Index in Hedonic Regression Models

S-1	S-2
0.129	0.148
(.000)	(.000)

TABLE 12.4
LM Test for Hedonic Regression Models

	S-1	S-2
LM_{lag}	99.69	123.77
	(.000)	(.000)
LM_{err}	68.90	91.28
	(.000)	(.000)

TABLE 12.5
Hedonic and Spatial Hedonic Models Estimated

Variable	S-2	S-2 SAR	S-2 SEM
(Intercept)	15.051	8.932	13.311
	(.000)	(.000)	(.000)
IMPROV	−0.170	−0.155	−0.155
	(.000)	(.000)	(.000)
ROOMS	0.274	0.262	0.263
	(.000)	(.000)	(.000)
TER	0.135	0.113	0.117
	(.000)	(.000)	(.000)
GAR	0.300	0.253	0.269
	(.000)	(.000)	(.000)
LIFT	0.292	0.260	0.281
	(.000)	(.000)	(.000)
POPSQM	−158.2	−90.80	−75.82
	(.000)	(.000)	(.046)
POPSQM^2	1813.9	1086.1	917.14
	(.000)	(.000)	(.038)
EMP	0.000	0.000	0.000
	(.000)	(.008)	(.468)
CBD	−0.001	−0.005	−0.005
	(.617)	(.014)	(.083)
TRANS	−0.221	−0.165	−0.169
	(.000)	(.000)	(.001)
CEN	0.374	0.266	0.018
	(.008)	(.042)	(.927)
COM	−0.019	−0.036	−0.089
	(.783)	(.582)	(.370)
GREEN	0.059	0.062	0.083
	(.368)	(.306)	(.352)
PG	0.251	0.077	0.238
	(.000)	(.242)	(.015)
$\rho\backslash\lambda$	–	0.382	0.527
		(.000)	(.000)
Log-likelihood	−296.2	−250.0	−260.3
Likelihood ratio test	–	92.29	71.68
		(.000)	(.000)

SAR and SEM models were estimated (S-2 SAR, S-2 SEM) using the variable journey time to the city centre by public transport in the specification (CBD) given that the models with this variable provided a better fit than those estimated using the gravity accessibility indicator in Example 10.1. The models had a goodness of fit, which was significantly better than conventional regression model (S-2) according to the LR test. If the parameters estimated using the spatial models are compared with those estimated with the conventional hedonic model, not that much of a difference is seen in most cases, though, for example, the estimated parameter for CBD was clearly not significant in the conventional hedonic model. Furthermore, it should be highlighted that with the SAR type of models, the independent variable does not only impact the local dependent variable directly but also the neighbouring observations (and from these onto the local observation) that need to be considered when studying the overall impact of each variable. Given that the correlation in the case of the SEM model is only produced at the level of the model's residuals, the estimated parameters can be directly interpreted as in a non-spatial regression model (LeSage and Pace 2010b). Therefore, for example, in the S-2 SEM model, each additional minute of journey time to the city centre using public transport implies a fall of 0.5% in the average house price.

Of the two models, the S-2 SAR provides the best fit and could be the best candidate for making predictions about house prices within a LUTI model. Another option could be to estimate a model that combines dependence in the dependent variable and the residuals specifying an identical or different neighbourhood matrices for both cases.

12.4 SPATIAL DEPENDENCE IN TRIP GENERATION MODELS

Various techniques are available to simulate how many journeys are produced and attracted by each zone in a study area (de Dios Ortúzar and Willumsen 2011). Linear regression is one of the more frequently used techniques for simulated trip generation and, as mentioned previously, is based on the hypothesis of independence in the model residuals.

Linear regression models applied to trip generation are normally estimated using household data. The parameters of the model will, therefore, provide trip generation rates for each of the variables being introduced. Some of the more frequently used variables are number of household members, number of vehicles owned by the household members and income levels, among others. The data are later expanded onto a zonal scale by introducing the average zonal values of each of the variables into the models and multiplying the trip generation rate by the total number of households in each zone. This allows the researcher to simulate the home-based trips produced in each zone in the study area.

However, this type of modelling may generate spatially dependent residuals. This spatial dependency between each household's trip production can be interpreted as a diffusion effect from similar mobility patterns between nearby households. Phenomena related to different social norms, transport habits and other factors can result in spatial dependencies in the trip generation behaviour of nearby households (Cordera et al. 2016). The errors can also present spatial dependency derived from the omission of some relevant variable from the model specification or from measuring the variable with a certain level of error. In either case the spatial extension of the variable could have a spatial differential effect, generating residuals correlated in certain areas. Both effects can be captured by using SAR and SEM models that consider this correlation in the residuals and thereby allow unbiased and/or efficient parameters to be estimated.

Example 12.3: Prediction of Trip Generation by Using Models Considering Spatial Dependence between Observations

Given the method for estimating and checking the validity of a regression model considering spatial dependence between observations is analogous to those presented in Example 12.2, we will examine in this particular case about how this type of model can be used for making predictions about future scenarios.

In the case of the SAR model, the prediction should consider the relationship of spatial dependence between the trips generated from neighbouring observations. Bivand (2002) proposed, in a similar way to time series models, the decomposition of the terms of the SAR model into three components: the trend, the signal and the noise (Equation 12.13).

$$\hat{y} = \underbrace{\hat{\rho}Wy}_{\text{Signal}} + \underbrace{X\hat{\beta}}_{\text{Trend}} + \underbrace{\varepsilon}_{\text{Noise}} \tag{12.13}$$

The trend is given by the term $X\beta$ in both the SAR and SEM models, in other words, by the matrix of the independent variables and their estimated parameters. In the SAR model, the signal is calculated using the following expression:

$$\hat{y} = (I - \hat{\rho}W)^{-1}X\hat{\beta} \tag{12.14}$$

where:
$\hat{\rho}$ is the estimated parameter of spatial autocorrelation
I is the identity matrix
W is the neighbourhood matrix

This approach carries the inconvenience of losing the part of the signal, which is incorporated into the error term. In the SEM model on the other hand, the signal is assumed to be equal to zero, meaning that the prediction of the model will only be given by the trend as in a conventional regression model. Finally, the noise is given by the stochastic term that is assumed to be independent and identical between observations.

The SEM model can therefore be used in a predictive mode like a conventional regression model. The values of the variables are substituted by each of their means in each of the zones and the overall trip generation will be the resulting average trip generation rate multiplied by the number of households located in the zone. Therefore, the total household journeys generated in zone i (T_i) could be given by

$$T_i = H_i(0,1 + 1,5\overline{X}_{1i} + 1\overline{X}_{2i}) \tag{12.15}$$

where:
$\overline{X}_{1i}$ is the average number of household members in each of the zones
$\overline{X}_{2i}$ is the average number of vehicles owned by each household in each zone
H_i is the number of households located in zone i

So, if zones 4 to 6 corresponding to Figure 8.5 (Chapter 8) have the characteristics reflected in Table 12.6, zone 4 would produce a total of 139,501 home-based trips, zone 5 would produce 45,941 trips and zone 6 would produce 37,790 trips.

TABLE 12.6

Characteristics of the Three Zones in the Example

Zone	Average Number of Persons per Household	Average Number of Vehicles per Household	Number of Households
4	2.5	1.5	26,075
5	2	1	11,205
6	3	1.5	6195

TABLE 12.7

Matrix Corresponding to the Signal of the SAR Model

$(I-\hat{\rho}W)^{-1}$	4	5	6
4	1.0	0.1	0.1
5	0.1	1.0	0.1
6	0.1	0.1	1.0

Nevertheless, with the SAR model, if we wish to calculate the total number of trips generated by each of the zones, then we need to consider the term $(I-\hat{\rho}W)^{-1}$, which generates autocorrelation in the dependent variable between neighbouring zones if the ρ parameter is significantly different from zero. Supposing a parameter ρ of 0.1, the signal of the model is given by the matrix presented in Table 12.7. With these values, zone 4 would produce a total of 154,312 home-based trips, zone 5 52,972 trips and zone 6 41,087 trips. The presence of spatial dependence between the zones therefore generates 25,139 extra trips. Obviously, a ρ value greater than 0.1 would result in the generation of even more trips in the area being studied.

12.5 CONCLUSION

Throughout this chapter, we have tried to underline the importance of considering the possible presence of spatial dependence effects in the data used for estimating different sub-models within a LUTI model. This spatial dependence may be commonly present in the cross-sectional data and the panel data, thereby making the parameters of the estimated regression or discrete choice models biased and/ or inefficient. It is suggested that models are estimated to explicitly consider spatial dependence in the data, where this occurs. This can be done by using SARs in the dependent variable or in the error term in the case of linear regression, and the nested, cross nested and EC models for the case of discrete choice.

When a significant level of spatial dependence is detected, the models that explicitly address it may provide more reliable parameters and more realistic substitution ratios between alternatives in the case of discrete choice models. These models will probably have a better goodness of fit to the data. It is therefore recommendable to explore the presence of situations of spatial dependence when estimating location, real estate price and regression sub-models among others in order to increase their reliability.

REFERENCES

Anselin, L. 1988a. Lagrange multiplier test diagnostics for spatial dependence and spatial heterogeneity. *Geographical Analysis* 20 (1): 1–17. doi:10.1111/j.1538-4632.1988. tb00159.x.

Anselin, L. 1988b. *Spatial Econometrics: Methods and Models, Studies in Operational Regional Science 4*. Dordrecht, the Netherlands: Kluwer Academic Publishers.

Anselin, L. 2001. Spatial econometrics. In *A Companion to Theoretical Econometrics*, B. Baltagi, (Ed.), pp. 310–330. Oxford, UK: Blackwell.

Anselin, L., and S. Rey. 1991. Properties of tests for spatial dependence in linear regression models. *Geographical Analysis* 23 (2): 112–131. doi:10.1111/j.1538-4632.1991. tb00228.x.

Anselin, L., and S. J. Rey. 2014. *Modern Spatial Econometrics in Practice: A Guide to GeoDa, GeoDaSpace and PySAL*. Chicago, IL: GeoDa Press.

Armstrong, R., and D. Rodríguez. 2006. An evaluation of the accessibility benefits of commuter rail in eastern Massachusetts using spatial hedonic price functions. *Transportation* 33 (1): 21–43. doi:10.1007/s11116-005-0949-x.

Ben-Akiva, M. E., and S. R. Lerman. 1985. *Discrete Choice Analysis: Theory and Application to Travel Demand*. Vol. 9, Cambridge, MA: MIT Press.

Bierlaire, M. 2003. Biogeme: A free package for the estimation of discrete choice models. *Proceedings of the 3rd Swiss Transportation Research Conference*, Ascona, Switzerland.

Bivand, R. 2002. Spatial econometrics functions in R: Classes and methods. *Journal of Geographical Systems* 4 (4): 405–421. doi:10.1007/s101090300096.

Bivand, R., M. Altman, L. Anselin, R. Assunção, O. Berke, A. Bernat, G. Blanchet, E. Blankmeyer, and M. Carvalho. 2015. Package 'spdep'. Spatial dependence: Weighting schemes, statistics and models. http://CRAN.R-project.org/package=spdep.

Brassel, K. E., and D. Reif. 1979. A procedure to generate Thiessen polygons. *Geographical Analysis* 11 (3): 289–303. doi:10.1111/j.1538-4632.1979.tb00695.x.

Cordera, R., Coppola, P., dell'Olio, L., Ibeas, Á. 2016. Is accessibility relevant in trip generation? Modelling the interaction between trip generation and accessibility taking into account spatial effects. *Transportation*, (In Press) 1–27.

de Dios Ortúzar, J., and L. G. Willumsen. 2011. *Modelling Transport*. Hoboken, NJ: John Wiley & Sons.

Fingleton, B. 2001. Theoretical economic geography and spatial econometrics: Dynamic perspectives. *Journal of Economic Geography* 1 (2): 201–225.

Fingleton, B. 2003. Externalities, economic geography, and spatial econometrics: Conceptual and modeling developments. *International Regional Science Review* 26 (2): 197–207.

Greene, W. H. 2012. *NLOGIT Version 5. Reference Guide*. Castle Hill, Australia: Econometric Software.

Guevara, C. A., and M. E. Ben-Akiva. 2013. Sampling of alternatives in multivariate extreme value (MEV) models. *Transportation Research Part B: Methodological* 48: 31–52. doi:10.1016/j.trb.2012.11.001.

Gujarati, D. N., and D. C. Porter. 2009. *Basic Econometrics*. 5th ed. Boston, MA: McGraw-Hill Irwin.

Hensher, D. A., J. M. Rose, and W. H. Greene. 2015. *Applied Choice Analysis*. 2nd ed. Cambridge, UK: Cambridge University Press.

Ibeas, A., R. Cordera, L. dell'Olio, P. Coppola, and A. Dominguez. 2012. Modelling transport and real-estate values interactions in urban systems. *Journal of Transport Geography* 24: 370–382. doi:10.1016/j.jtrangeo.2012.04.012.

Lee, B. H. Y., and P. Waddell. 2010. Residential mobility and location choice: A nested logit model with sampling of alternatives. *Transportation* 37 (4): 587–601. doi:10.1007/s11116-010-9270-4.

LeSage, J. P., and R. K. Pace. 2009. *Introduction to Spatial Econometrics, Statistics, Textbooks and Monographs*. Boca Raton, FL: CRC Press.

LeSage, J. P., and R. K. Pace. 2010a. The biggest myth in spatial econometrics. http://ssrn.com/abstract=1725503.

LeSage, J. P., and R. K. Pace. 2010b. Spatial econometric models. In *Handbook of Applied Spatial Analysis: Software Tools, Methods and Applications*, M. M. Fischer and A. Getis (Eds.), pp. 355–376. Berlin, Germany: Springer.

Long, F., A. Páez, and S. Farber. 2007. Spatial effects in hedonic price estimation: A case study in the city of Toronto. Working paper series, Centre for Spatial Analysis, Canada.

Moran, P. A. P. 1948. The interpretation of statistical maps. *Journal of the Royal Statistical Society. Series B (Methodological)* 10 (2): 243–251.

Tobler, W. R. 1970. A computer movie simulating urban growth in the Detroit region. *Economic Geography* 46: 234–240. doi:10.2307/143141.

Ward, M. D., and K. S. Gleditsch. 2008. *Spatial Regression Models, Quantitative Applications in the Social Sciences*. Thousand Oaks, CA: Sage Publications.

13 Practical Application of a Land Use–Transport Interaction Model

Juan Benavente, Alvaro Landeras,
Rubén Cordera and Ángel Ibeas

CONTENTS

This chapter will describe the application of the sub-models described in the book to calibrate a combined land use and transport interaction model. The steps taken to define the goals of the model, outline the study area and collect the data that are required as the input information for the sub-models were summarised in Chapter 8. The examples were presented in Chapters 9 through 11 showing how to estimate models for the location of population and economic activities, as well as real estate pricing and mobility simulation models (trip generation, trip distribution, modal choice and assignment). This chapter will explain how to interrelate these sub-models and how to perform a joined-up calibration of the entire land use–transport interaction (LUTI) system. The model was programmed using Python 2.7, a free code language compatible with a variety of transport modelling software, including the Visum® model, which was used in this case (PTV 2014). The resulting LUTI model has been applied to a real case in order to simulate the effects of introducing a specific transport policy. A LUTI model that considers spatial dependence in the data will also be calibrated and its fit will be compared with that of the standard model. The aim of these examples is to show how different models based on different theories can be integrated into a unified system, which can be applied to support decision-making in practical cases of urban and transport planning.

13.1   THE STRUCTURE OF THE LUTI MODEL

Figure 13.1 introduces the structure of the LUTI model, which is applied throughout this chapter. The blocks on the left of the figure represent the transport sub-model (demand models and assignment model), whereas the blocks on the right correspond to the land use sub-models: residential location, economic activities location and real estate pricing sub-models.

The system has a parsimonious structure, similar to that of a classic LUTI model (e.g. the Lowry model). However, differently from the classic systems, this LUTI model includes a real estate pricing sub-model as well as techniques for simulating the processes that are involved with the residential and economic activities location through random utility theory (see Chapters 6 and 9). The LUTI model presented here is therefore equipped with a wider theoretical foundation as it can be supported by an economic model addressing the behaviour of urban agents (companies, households…) based on the maximum utility. In terms of the interaction between the different sub-systems, the proposed model can be characterised by a comparative equilibrium. Any change occurring in any element of the territorial system (e.g. the introduction of new transport infrastructure) leads to an equilibrium solution for the system, in accordance with the theoretical hypotheses. The comparative equilibrium approach does not allow estimations to be made for the territorial system adjustment process, but it is an easier method to apply and is valid for a model of

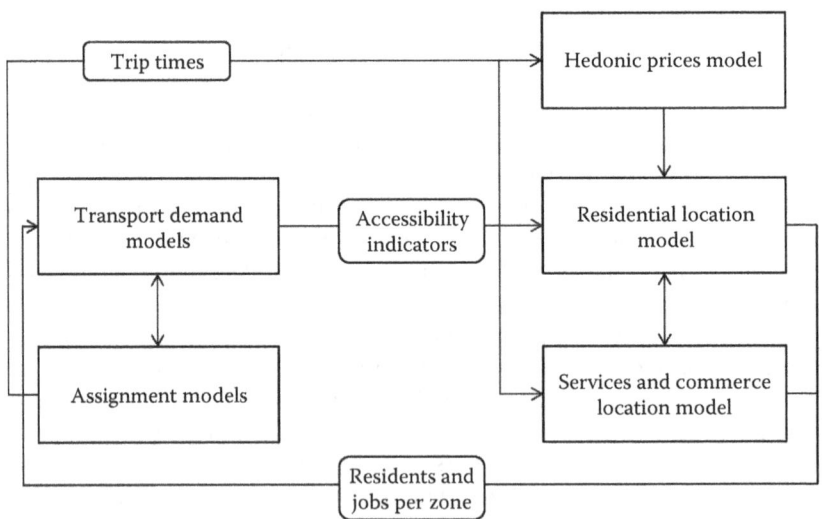

FIGURE 13.1   Structure of the proposed LUTI model.

the calibre presented here. More advanced systems based on the behaviour without an overall equilibrium involve considerable problems in their practical application because of their increased demands for data and their greater functional complexity. The application of such models is only viable in urban areas with a large amount of available data (Duthie et al. 2007), although with the growing availability of digitalised data, this problem will probably be considerably reduced in the near future.

The multiple interactions can be established between the different sub-models depending on the proposed theoretical hypotheses. First, it is logical to assume that the demand for transport in the different zones depends on the location of residents and economic activities within the study area. This demand for transport is also associated with a series of journey choices which, when related to the supply of the available transport (road capacities, available public transport services), will generate an inter-zonal cost matrix. This cost matrix can be expressed in terms of journey time, generalised transport costs by mode or a composite cost grouping all the available modes together, and will influence the accessibility levels of each zone.

The residential location sub-model (see Chapter 9) is based on the hypothesis that workers choose their zone of residence after considering a variety of local characteristics, such as distance to areas of employment represented by journey times and/or composite cost, which are provided by the interaction between the transport and economic activities location sub-models. Another basic variable affecting the residential location is the house prices in each zone, which, in turn, is provided by the real estate pricing sub-model.

The activities location sub-model behaves in a similar way to the residential location sub-model by considering the utility of each of the zones according to different variables, including the population accessibility level, which depends on the inter-zonal journey costs, derived from the transport model, and the population distribution pattern, which can be derived from the residential location model.

The implicit prices model can calculate the average house prices in each zone according to the structural and environmental characteristics of the area. The supply of transport can also be considered (e.g. public transport provision) along with other variables, which are theoretically relevant and usually introduce a political context into the research performed using a LUTI model.

There is a profound interdependence between the different sub-models, an interdependence that can be increased or decreased by modifying the complexity of the functional specifications. Without doubt, this could increase the model's realism, although this would come at the cost of also increasing the errors resulting from the inclusion of more variables (which come with an associated measurement error) and an increase in the costs of running the model (de Dios Ortúzar and Willumsen 2011). Furthermore, the circular dependence between the different sub-models generates the problem of finding an equilibrium, which will be addressed later.

The LUTI model can also consider the presence of a constraint on the capacity of the zone in terms of accepting both population and economic activities. These limits are set exogenously and help to improve the goodness of fit and the realism in the model's simulations. The constraints on capacity in a future scenario should be set considering the urban and the territorial planning within the study area, aiming not to be too restrictive, so as not to reduce the flexibility of the simulated results.

Finally, the model starts from a series of theoretical hypotheses, which help to simplify the complex territorial reality by excluding the aspects considered to be secondary for obtaining relevant results from the simulation. For purposes of illustration, in the model presented here these hypotheses can be summarised as follows:

1. The study area being modelled is considered to be closed; in other words, the jobs are occupied by the internal demand. This closure of the territorial system allows the process to remove problems that result from workforce immigration/emigration. This hypothesis requires the modelled area to have a closed nature in terms of daily journeys, so a metropolitan character is recommended, as detailed in Chapter 8.
2. The land market is outside the scope of the model, so no mechanism of residential supply and demand equilibrium is presented. A real estate supply sub-model will not be estimated, so the resulting housing prices maintain the base year equilibrium in the housing market.
3. The location of economic activities belonging to the basic sector is outside the scope of the model. Such activities locate without considering the population distribution in the study area as they do not directly depend on the internal demand. The distribution of this kind of employment can therefore be considered as fixed or only modified by factors outside the workings of the model.

13.2 CONNECTION AND EQUILIBRIUM BETWEEN LAND USE AND TRANSPORT MODELS

The land use and transport sub-models interact by trip generation in the case of the relationship land use → transport and through accessibility indicators and journey costs between zones in the case of the relationship transport → land use.

The land use → transport connection: Trip generation, considering both the production and attraction of journeys, is influenced by the characteristics of the zone in question, such as the population and the availability of jobs. As explained in Chapter 11, the trip generation model, as the first phase of a transport model, should be estimated considering these zonal variables to take in the fact that in a LUTI model the population and jobs present in each zone are endogenous variables.

The transport → land use connection: The connection between transport and land use depends on the indicators of journey costs between the zones. These journey costs can consider only a single variable as being representative of the total cost (e.g. on board journey time), the overall journey cost considering all the variables involved (journey time, waiting time, cruising time for parking, etc.) or the composite cost formed by considering all the modes of transport that are available in a determined zone. Of all these possibilities, the composite cost is closest to the reality as it consistently aggregates the costs of all the available transport modes in the zone.

The different sub-models can be run iteratively until a final solution in equilibrium is reached. The analyst should use an input parameter to specify a maximum change value in the location pattern and in the transport flows between the consecutive iterations to evaluate if the model has generated a solution in equilibrium that represents the final state of the territorial model.

There are two large problems of equilibrium in the proposed LUTI model. First, because the location of the population depends on the location of economic activities and vice versa, the following problem appears:

$$\begin{cases} R^i = R\left[\sum_i A^i\right] & \forall i \\ A^i = A\left[\sum_i R^i\right] & \forall i \end{cases} \tag{13.1}$$

where:
  R^i is a vector [n_zones × 1] of type i residents
  A^i is a vector [n_zones × 1] of type i jobs

The solution to this equilibrium problem can be approached as a fixed point problem in which the solution is given by the vectors R^{i*} and A^{i*}. The existence of at least one solution is guaranteed by Brouwer's fixed-point theorem (Cascetta 2009), which states that the functions R and A are to be continuous and that the domain of the functions be a not empty, compact (closed and bounded) and convex set. Both conditions are fulfilled if the locations of population and activities are simulated using a logit type discrete choice model.

An equilibrium problem is also associated with the pattern of population and activities location and the journey pattern provided by the transport model from the trip matrix. According to the chosen modelling system, both sub-systems can be approached as separate equilibriums or as a joined equilibrium. In the example presented in the following section, the equilibriums of both sub-systems are addressed separately. Whereas the transport model can implement a classic style demand/supply equilibrium (see Chapter 11), the equilibrium of the LUTI system is assumed to be reached when the journey costs provided by the transport model do not induce changes in the location patterns of the population and employment. A simple iterative algorithm can be used, one that can easily reach a solution within few iterations, if and when the functional structure of the model does not present a high level of complexity.

13.3 IMPLEMENTING A LAND USE–TRANSPORT INTERACTION MODEL

As highlighted in the introduction, the LUTI model presented was run using a combination of commercial software in the case of the transport model and a specific software developed ad hoc for the location and real estate pricing models. The latter case used the Python 2.7 programming language given its compatibility with the transport model, its flexibility and its acceptable computing performance compared with other languages (Aruoba and Fernández-Villaverde 2014).

Python is an interpreted programming language. The interpreter directly translates each line of code as it is being executed, meaning that it can work on multiple platforms through its virtual machine. Furthermore, it is a multi-paradigm language, which incorporates elements of both functional and procedural programming (Lutz 2013).

The software has been designed in a main block that runs iteratively until the location pattern is stable. The various associated functions also allow different calculations to be made as required by the model: aggregation and disaggregation between transport zoning and the land use zoning data, calculation of regression models, calculation of utilities, calculation of location probabilities and calculation of basic accessibility indicators.

13.3.1 CALIBRATION OF THE JOINED-UP MODEL WITHOUT CONSIDERING SPATIAL DEPENDENCY BETWEEN THE OBSERVATIONS

Before being able to use a LUTI model for assessing policies and projects, its goodness of fit with the base year calibration data needs to be considered to make the results as comparable as possible with the current situation. In this case, the available data correspond to the year 2008, which will therefore be taken as the corresponding base year.

The model took 4 iterations to reach a solution in equilibrium. The stop criteria was that the difference between the population and activities location results between the consecutive iterations must be less than 1% for the model to have found a solution. The fit was improved by setting similar occupancy restrictions on population and jobs to those found in consolidated zones and up to 10% greater than in the peripheral areas where new activities and populations could be established.

The residential location model as a whole, aggregating the data referring to people with medium–low and high incomes showed a goodness of fit with the base year observed population of 0.94 using the R^2 indicator (see Figure 13.2a). The model managed to capture the population location pattern with a mean absolute percentage error (%MAE) of 4%. The %MAE is calculated using

$$\%\text{MAE} = \frac{\sum_{i}^{n} \frac{\left|P_{\text{obs}} - P_{\text{mod}}\right|}{P_{\text{obs}}}}{n} \tag{13.2}$$

where:
P_{obs} is the observed magnitude (e.g. population)
P_{mod} is the modelled magnitude
n is the number of observations

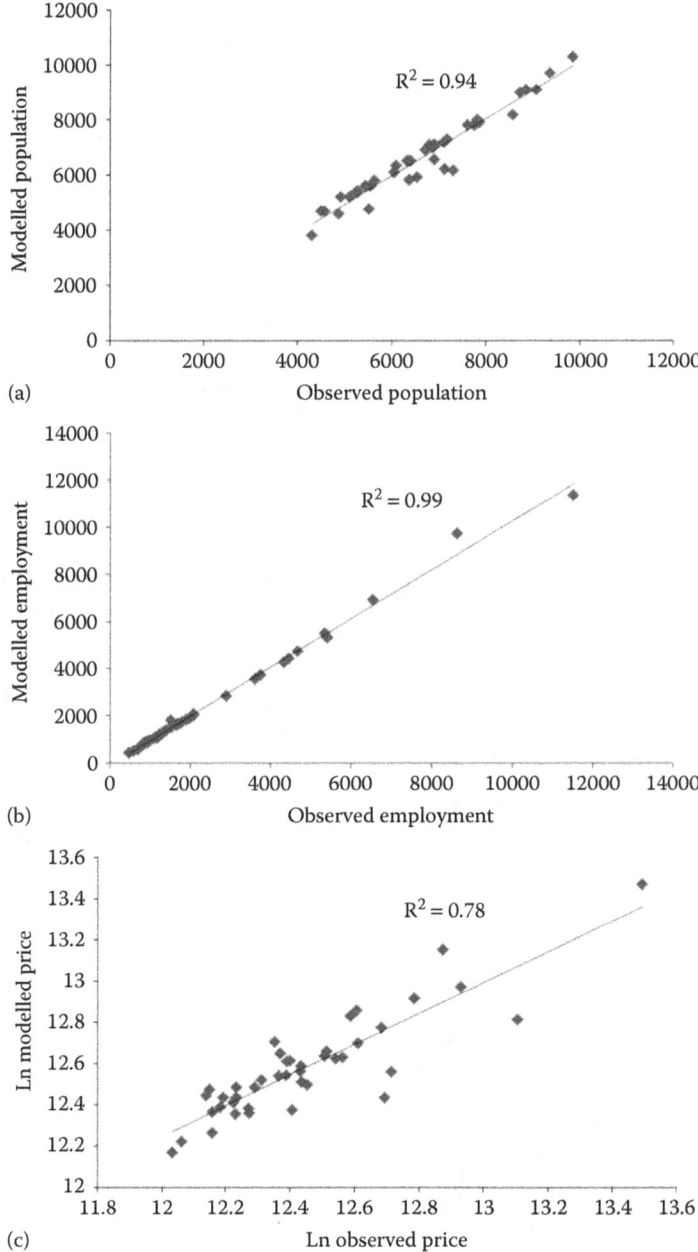

FIGURE 13.2 Fit between the observed data and calibrated data for (a) population, (b) economic activities and (c) real estate prices.

Another indicator that could also be used to test the fit of the model is the mean square error (MSE), which strengthens the deviations to a greater degree over the observed magnitudes, as

$$\text{MSE} = \frac{\sum_i^n \left(P_{\text{obs}} - P_{\text{mod}}\right)^2}{n} \qquad (13.3)$$

Grouping together all the economic sectors, the fit of the activities location model showed an R^2 of 0.99 (see Figure 13.2b) and generally represented by the location pattern of economic activities quite precisely with a %MAE of 4%. Finally, the goodness of fit of the real estate prices model had an R^2 of 0.78 (see Figure 13.2c). Most of the errors were found in the more distant zones from the city centre where the model tended to overvalue the average house prices. Overall, the %MAE of the real estate pricing model was around 1%.

Figure 13.3 shows the traffic assignment from the transport model, derived from the location of population and employment adjusted to the base year. Most vehicles are concentrated on the northern and southern access routes to the urban centre where the majority of the commercial premises and services are located. The volume of vehicles at peak times on these higher capacity roads can be more than 3000 per hour according to the estimations made by the model.

If the results produced by the model do not fit well enough with the observed base year data, then a step-by-step review of the parameters estimated in each of the sub-models is required to check for errors in the functional specifications or the measurement of the variables. It is also important to check that the values of the input data for each sub-model, coming from both endogenous and exogenous data from other sub-models, are within the expected margins of error. Finally, specific parameters for individual zones can be introduced to improve the model's fit, particularly where

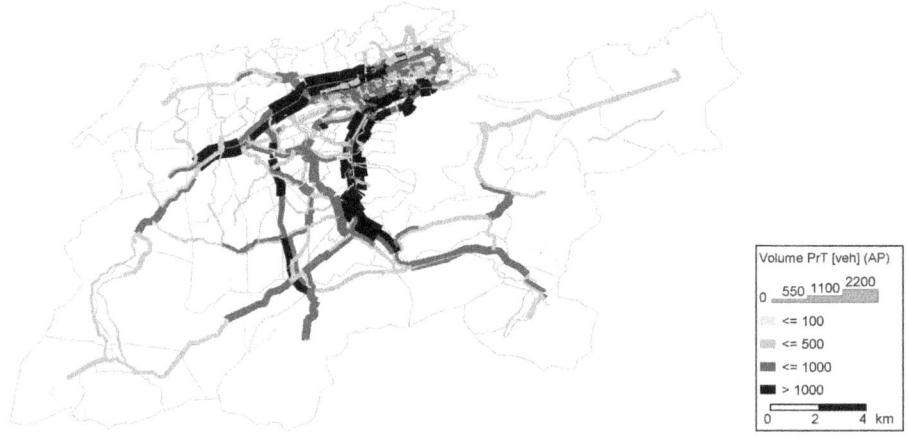

FIGURE 13.3 Traffic assignment for private vehicles on the transport network.

this is lower than expected. However, the modeller also needs to consider that a greater number of ad hoc parameters reduce the model's flexibility when simulating future scenarios.

13.3.2 USING A LAND USE–TRANSPORT INTERACTION MODEL TO SIMULATE THE INTRODUCTION OF A NEW POLICY: CHARGE FOR ACCESSING THE URBAN CENTRE

Once the model has been calibrated and validated to have an acceptable fit with the base year data, it can be used to make predictions derived from the introduction of new policies and projects, or to observe the trends resulting from changes made to the input variables. To illustrate the possibilities in the study area, the calibrated model will be used to evaluate the medium-term implications of introducing a charging policy for accessing the city centre by private vehicles. The results of this policy will be compared with the results of the base year calibration to examine if this would really have an impact on the distribution patterns for population, economic activities and mobility in the study area. This kind of charging policy has already been introduced in large cities such as London, Singapore and Stockholm with generally positive results for the change towards the sustainable mobility (Santos and Shaffer 2004, Eliasson 2009).

To present the repercussions of these changes in the urban system, the 42 zones presented in Chapter 8 will be grouped into 5 large zones in order to improve the readability of the results and more directly capture the changing patterns experienced across the territorial system. These five large macro-areas have been defined as shown in Table 13.1. The charging policy will be applied on the access routes to macro-area 1. This charge was calculated as a payment of €4 to take a car into the central zone. This central zone is where the highest demand for transport is located due to the higher presence of commercial activities and services.

When the change is applied to the transport model by increasing the cost on the entrance links to the charging zone, the model needs to be run again until a solution between the transport and the location models is reached. In this case, a solution was reached after 5 complete iterations, once again, after applying a stop criteria of a lower than 1% difference between the two consecutive iterations in the location patterns of population and activities.

TABLE 13.1
Grouping of Land Use Zones into Macro-Areas

Name of Macro-Area	Zones Included
1. Central Santander	1 and 2
2. Suburbs Santander	3 to 8 and 11 to 21 and 23, 24
3. Periphery Municipal Santander	9, 10, 22, 25 and 26
4. First Metropolitan Ring	27 to 35
5. Second Metropolitan Ring	36 to 42

In terms of mobility, the model showed the introduction of charges resulting in a drop between 4% and 7% in the trip attraction of the affected zones. The assignment also showed some reduction in traffic flows, especially in the area to the west of the new charging zone (see Figure 13.4 in which the dotted line outlines the transport zones inside the charging zone). However, the results did not show an increase in the use of public transport in the overall study area; on the contrary, its weight in the modal distribution stayed the same with respect to the results from the base year with 9% of the total trips made.

In order to improve the reliability of the predictions made about the effect of introducing this policy on the various location patterns, a pivot point technique can be applied (Manheim 1979, Coppola and Nuzzolo 2011) using the following correction factor:

(a)

(b)

FIGURE 13.4 Results of the private transport assignment (a) before and (b) after the application of charging to access the city centre. The dotted line outlines the zones included in the charging zone.

$$A(o) = \left(\frac{A_{census}}{A_{ref}} \right) A_{future} \tag{13.4}$$

where:

$A(o)$ is the predicted number of urban agents (population or jobs) in the already pivoted zone

A_{census} is the observed number of agents

A_{ref} is the number of agents predicted by the model in the reference scenario

A_{future} is the number of agents predicted by the model in the proposed scenario without applying pivot point technique

The application of this factor to the future data improves the precision of the model results by fitting better to the base year data.

After applying the pivot point technique the LUTI model provides the percentages of change presented in Table 13.2. Generally, it can be stated that introducing a charging system for taking a car into the central zone would result in a medium-term change in the location of population and employment. The city centre and the areas forming part of the urban nucleus of Santander (zones 1, 2 and 3) would lose population as opposed to the more outlying zones and the city's sphere of influence (zones 4 and 5), which would receive population that has displaced from the urban nucleus. Employment, on the other hand, would experience growth in zone 1 derived from the improved accessibility coming from the fall in transport costs inside the charging area. However, zone 2 would also see a fall in its level of activity, whereas zone 3, more peripheral but still not too far from the city centre, would see an increase in the number of jobs. House prices, however, would not see any significant change.

In terms of accessibility, the simulation shows that the introduction of a charging system (Figure 13.5) would see an increase in the population accessibility of zone 1, given the reductions in travel time between the internal zones. In zone 3, however, the population would experience reduced accessibility due to the new charge and the displacement of population to zones 1 and 3, which explains the drop in employment.

TABLE 13.2

Percentage Change in Location Patterns and Housing Prices in the Simulated Area Compared with the Base Year

Macro-Area	Population Change	Employment Change	Housing Price Change
1	−3.7	4.8	0.0
2	−0.8	−2.8	0.1
3	−3.5	2.2	0.0
4	2.0	0.0	0.0
5	4.2	−3.7	−0.1

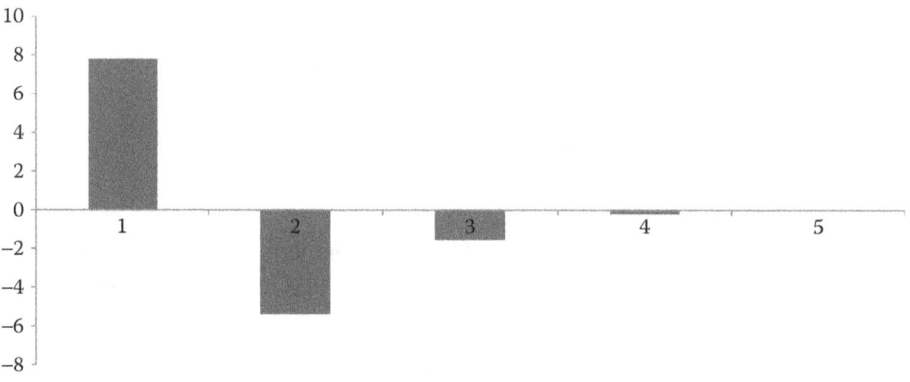

FIGURE 13.5 Percentage change in accessibility levels for the population in the macro-areas.

A charging policy in which the fee is too high for accessing the main areas of opportunities could therefore result in population dispersion and, to a lesser extent, the movement of activities away from the urban centre. In the case of the study area considered in this example, this process has already begun, with a fall in population of almost 7% over the past 20 years from the urban centre moving to the outlying areas within the city's sphere of influence. The model indicates that this dispersion process could be strengthened by introducing congestion charging in the city centre; however, this would not cause a reduction in the area's economic weight because of the improved mobility within the charging zone. Greater population dispersion can have a negative effect on the sustainability because residents need to use their private cars more often for basic activities and public transport services need to cover more extensive areas with lower population densities.

13.4 CALIBRATION OF A LAND USE–TRANSPORT INTERACTION MODEL CONSIDERING SPATIAL DEPENDENCE

A similar LUTI model to the one calibrated in Section 13.3 will be presented with the additional consideration of spatial dependence in both the real estate pricing model and the residential location model. Spatial dependence in the case of the real estate pricing model has been addressed using a spatial autoregressive model (SAR model) with spatial autocorrelation, whereas dependence between choice alternatives has been implemented using a nested logit model. Both are identical to the models estimated in Chapter 12.

The consideration of spatial dependence in the data used to calibrate the model should improve its level of fit and provide more reliable parameters. In this case, the availability of data has reduced the study area to 26 zones covering only the municipal borough of Santander.

The LUTI model as a whole was calibrated once again to test if a better goodness of fit was found than with the model, which did not consider spatial dependence

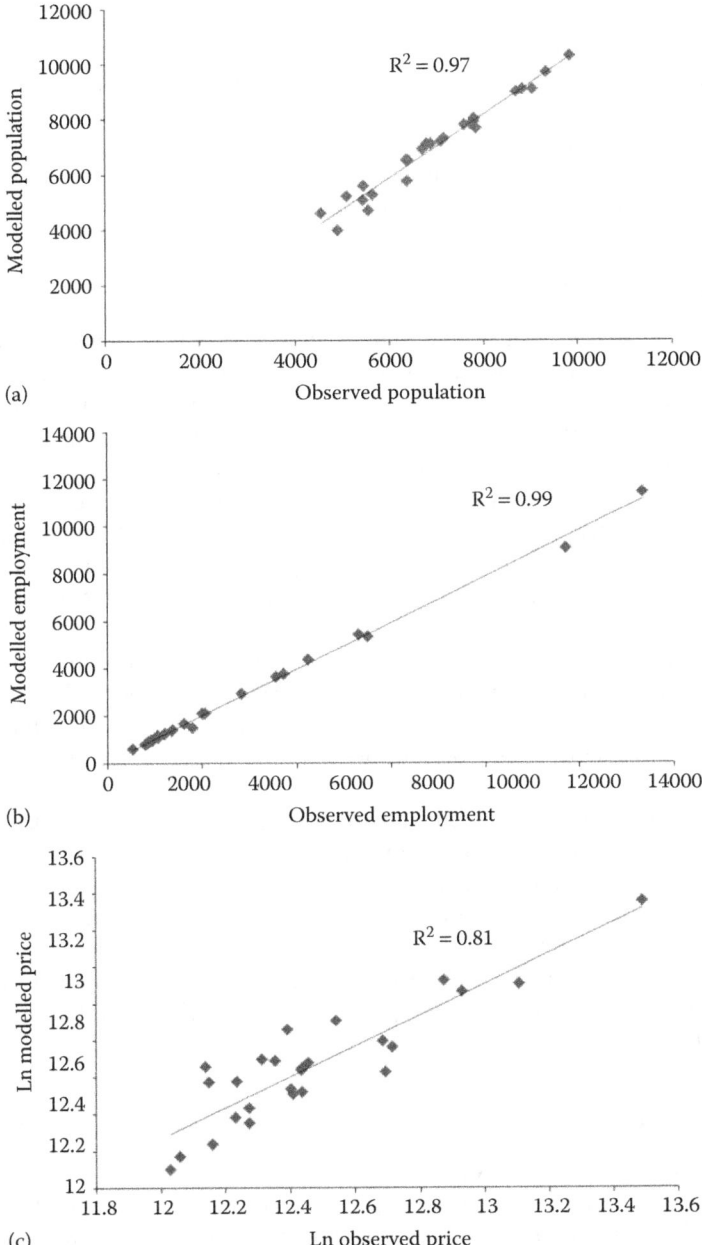

FIGURE 13.6 Fit between the observed and calibrated data for (a) population, (b) economic activities and (c) housing prices in the LUTI model considering spatial dependence.

between choice alternatives and between housing prices. After 3 iterations, the model converged on a solution.

Figure 13.6 shows the fits with the observed population, employment and house price data. The fits for the population model and the real estate pricing model were slightly better than with the previously calibrated LUTI model. The functional form of the activities location model was not altered and its level of fit was identical to that of the LUTI model that did not consider spatial dependence.

13.5 CONCLUSIONS

This chapter has described the overall structure of a LUTI model, combining different transport and land use sub-models, as well as testing its goodness of fit to the data observed in the base year. It has also been shown how a LUTI model can be applied to simulate the impact of introducing new policies or projects on an urban system. Interpreting the results of the LUTI model can provide relevant information to evaluate whether or not the effects of introducing different measures will be as expected. These effects can sometimes be surprising and are difficult to capture using traditional planning methods without the support of formal models because of the complexity involved in the interrelationships between the diverse elements making up the urban system.

The model estimated in the study case had a high goodness of fit to the data, and the introduction of modelling techniques to consider spatial dependence was able to further improve this fit. Spatial econometric techniques are able to help control the presence of effects that cannot be considered by the models based on more restrictive hypotheses.

However, it is important to remember that the predictions made by LUTI models, even in their more sophisticated versions, are based on a simplification of reality. The estimated parameters and the specifications of the different functions should be subjected to a continual process of review as new data becomes available and any predictions need to be carefully interpreted by the modeller. Nevertheless, LUTI models provide a great support tool in decision-making processes by helping us understand future scenarios for urban and transport planning clearly supported by theory and empirical evidence.

REFERENCES

Aruoba, S. B., and J. Fernández-Villaverde. 2014. *A Comparison of Programming Languages in Economics*. Cambridge, MA: National Bureau of Economic Research.

Cascetta, E. 2009. *Transportation Systems Analysis: Models and Applications*. 2nd ed, Springer optimization and its applications. New York: Springer.

Coppola, P., and A. Nuzzolo. 2011. Changing accessibility, dwelling price and the spatial distribution of socio-economic activities. *Research in Transportation Economics* 31 (1): 63–71. doi:10.1016/j.retrec.2010.11.009.

de Dios Ortúzar, J., and L. G. Willumsen. 2011. *Modelling Transport*. New York: John Wiley & Sons.

Duthie, J., K. Kockelman, V. Valsaraj, and B. Zhou. 2007. Applications of integrated models of land use and transport: A comparison of ITLUP and URBANSIM land use models. *54th Annual North American Meetings of the Regional Science Association International*, Savannah, GA.

Eliasson, J. 2009. A cost–benefit analysis of the Stockholm congestion charging system. *Transportation Research Part A: Policy and Practice* 43 (4): 468–480. doi:10.1016/j.tra.2008.11.014.

Lutz, M. 2013. *Learning Python*. Sebastopol, CA: O'Reilly Media.

Manheim, M. L. 1979. *Fundamentals of Transportation Systems Analysis; Volume 1: Basic Concepts*. Cambridge, MA: MIT Press.

PTV AG. 2014. *VISUM 14 User Manual*. Karlsruhe, Germany: PTV Group.

Santos, G., and B. Shaffer. 2004. Preliminary results of the London congestion charging scheme. *Public Works Management & Policy* 9 (2): 164–181. doi:10.1177/1087724x04268569.

Index

Note: Page numbers followed by f and t refer to figures and tables respectively.

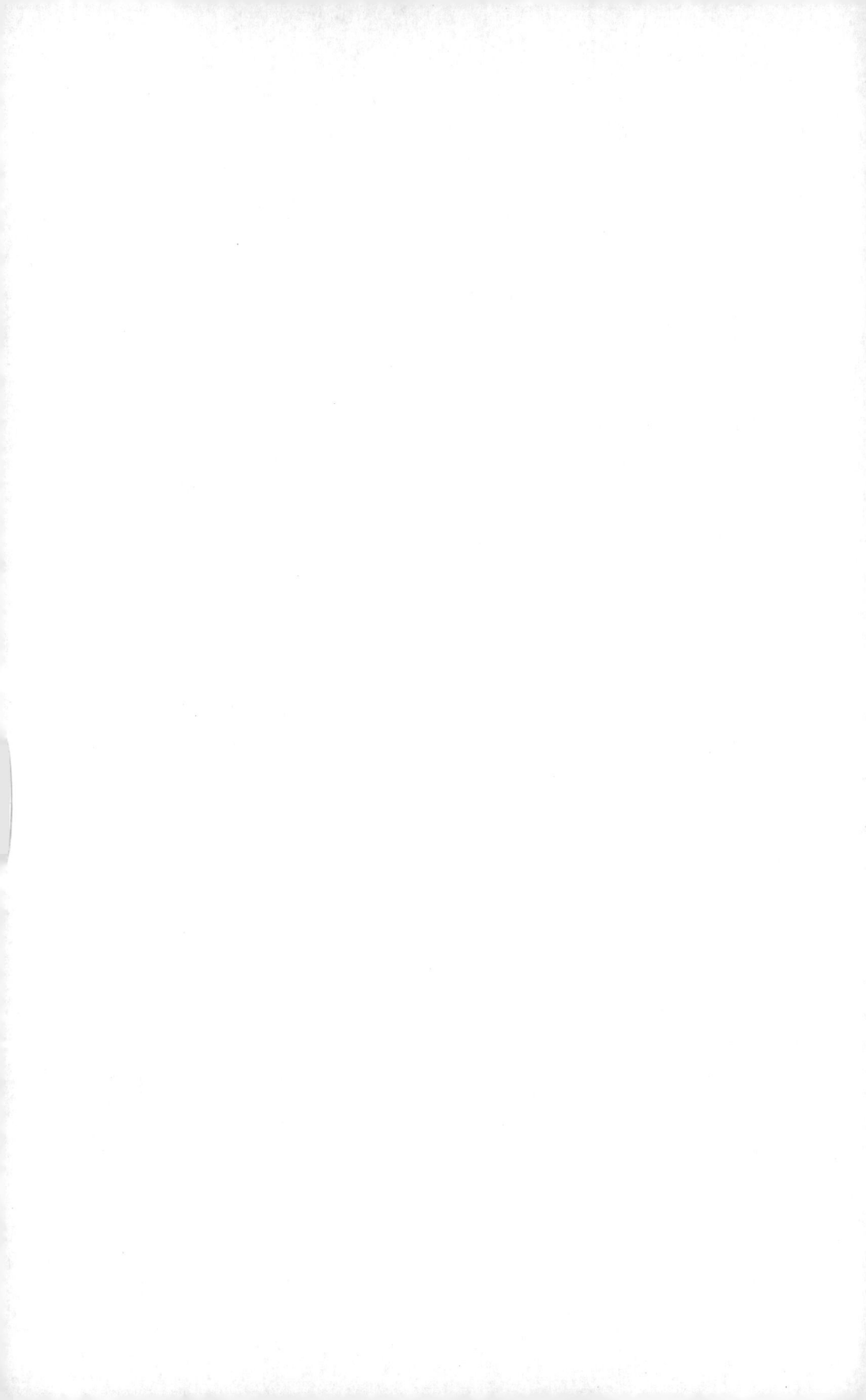